Burning
Issues

SUSTAINABILITY

AND MANAGEMENT

OF AUSTRALIA'S

SOUTHERN FORESTS

Mark Adams, Bushfire CRC, The University of Sydney,
and Peter Attiwill, The University of Melbourne

CSIRO

PUBLISHING

bushfire CRC

National Library of Australia Cataloguing-in-Publication entry

Adams, Mark A. (Mark Andrew)

Burning issues : sustainability and management of Australia's southern forests/ by Mark Adams and Peter Attiwill.

9780643094437 (pbk.)
9780643103467 (epdf)
9780643103474 (epub)

Includes bibliographical references and index.

Forest fires – Australia, Southern – Prevention and control.
Forest ecology – Australia, Southern.
Sustainable development – Australia, Southern.

Attiwill, P. M.

634.96180994

Published by
CSIRO PUBLISHING
36 Gardiner Road, Clayton VIC 3168
Private Bag 10, Clayton South VIC 3169
Australia

Telephone: [+613] 9545 8555
Local call: 1300 788 000 (Australia only)
Fax: +61 3 9662 7555
Email: csiropublishing@csiro.au
Website: www.publishing.csiro.au

Front cover: photo by Jason South/Fairfax Photos

Set in 10/12 Adobe Minion and Stone Sans
Edited by Peter Storer Editorial Services
Cover and text design by James Kelly
Typeset by Desktop Concepts P/L, Melbourne
Index by Russell Brooks
Printed by Ingram Lightning Source

CSIRO PUBLISHING publishes and distributes scientific, technical and health science books and journals from Australia to a worldwide audience and conducts these activities autonomously from the research activities of the Commonwealth Scientific and Industrial Research Organisation (CSIRO).

Foreword

Southern Australia is one of the three most fire-prone areas on Earth. After more than a century of urban growth and valiant efforts to 'tame' the bush, recent decades have seen more and more people moving back onto the fringe, or into the middle, of this volatile landscape.

As this movement has intensified so has the debate on how to best protect life and property from the ever-present bushfire threat. A long-running drought and a predicted warming climate have ensured that bushfire is a dominant factor in our nation's long-term planning.

Following the tragic Victorian Black Saturday fires in 2009, a much greater urgency now confronts policy makers, land and fire managers and communities living in bushfire areas. This has lead to a call for a single, simple answer on fuel-reduction burning to reduce the bushfire risk. *Burning Issues* explains that this is a complex issue without such a simple answer.

Fire has been an integral part of the Australian environment for millions of years, largely shaping the composition and distribution of the native plants, animals and ecosystems that survive today. Indeed, a significant proportion of Australia's unique biota has become largely dependent on fire for its continued existence and development.

The arrival of people into this environment, more than 40 000 years ago, saw new sources of fire other than lightning become a feature of our environment.

With later new arrivals to the southern part of this continent, infrastructure was introduced as a feature of the environment – farms, houses and industry. Uncontrolled fire came to be seen solely as a destroyer of life and property. This annual threat of bushfire has understandably led to a focus on protecting human lives and assets.

The Bushfire Cooperative Research Centre was established in 2003. Its formation represented a major commitment by the fire and land management agencies and their research partners in Australia and New Zealand towards a better, collective understanding of the complex social, economic and environmental issues associated with bushfire.

The combination of partner resources and an Australian Government grant, through its CRC Program, has seen a substantial investment in research that has produced findings of national and international significance. This commitment to ensure that key decisions on bushfire management are based on sound research will continue as new issues emerge.

From the outset, the fire and land management agency partners in the Bushfire CRC, and the Australasian Fire and Emergency Service Authorities Council, identified fuel reduction as one of the most important issues confronting them, both in a technical and scientific sense, and in terms of community perception and understanding of bushfire control and management.

'Managing Fire in the Landscape' was one of the four major research programs that were established to guide the Bushfire CRC's work. That program, in conjunction with complementary research being undertaken elsewhere by the Bushfire CRC, was designed to assist fire and land managers and the broader community to better manage fire and to understand its importance as a land management tool.

Across southern Australia, a key component of any sound bushfire preparedness strategy is the management of fuel on selected areas through 'cool' burning. Fuel-reduction burning is a powerful strategic tool for both the bushfire manager and the ecosystem manager.

A successful program of fuel-reduction burning is sometimes difficult to achieve given the limitations in weather forecasting, the uncertainty of fuel moisture predictions, the availability of fire managers with the right skills and sufficient funding. The approach taken by most agencies to fuel-reduction burning is necessarily selective and strategic.

In the more densely settled areas of Australia in particular, it is not well understood that the best way to protect life and property from bushfires is to manage and use fire in the environment. Attempts to simplify the debate are often problematic – both from a perspective of risks to life and property and in terms of the needs of our plants, animals, soils and water.

Burning Issues brings together a considerable amount of the work that has been undertaken over the last 50 years. It provides a synthesis of our current knowledge of fire and its interactions with the environment and it examines the evidence on the management of the risks to life and property.

The Bushfire CRC acknowledges that there are other views on this controversial topic. However, the publication of this book is done with the aim of moving the debate from the extremes, providing policy makers, land and fire managers and rural communities with a comprehensive, scientific knowledge base. Thoughtful use of the book's contents should encourage confident judgements about the management of risk and the maintenance of environmental values in a region.

Clearly, the debate must start from the objectives for the management of the particular land in question, and a careful consideration of the nature of the risks. Only then will an understanding of the 'appropriate' levels of fuel-reduction burning emerge.

This book sets out to ensure that our ongoing discussions about the management of land and the associated risk are as informed as possible.

Finally, and on behalf of the Board of the Bushfire CRC, I would like to thank the book's authors and indeed all the researchers who have contributed so constructively and so rigorously to our collective understanding of the management of our fire-prone landscapes.

Len Foster AO
Chairman
Bushfire CRC

Contents

Preface

This book began as a text on planned fire. It still is, but as it developed and as the issues mounted, it became clear that a broader context or framework was necessary. The broader framework is that of 'ecological sustainability'. We have used here an ecological definition of sustainability because it provides a sharper focus on the key science issues, rather than the often-used and broader form of definition that includes issues of social and inter-generational equity.

Fire is crucial to any effort to achieve ecological sustainability in Australia and will become more so as the climate warms and dries. Although concerned scientists have been aware of the essential role of fire for decades, many among the general public and policy makers at all levels of government remain unaware just how crucial fire regimes are to the critical components of ecological sustainability: in maintaining biodiversity, for soil development and in the carbon, water and nutrient cycles. Of these, only biodiversity has been given anything like sufficient attention and, even then, it has been attention that has done more to mystify than to clarify.

Other elements of ecological sustainability are now widely debated. This is partly because water has an obvious dollar value, as does carbon, and partly because, finally, the general public and policy makers are slowly becoming aware that water and carbon are important to the future of the planet. On this, the world's driest continent (apart from Antarctica), fire – much more than showering in buckets or planting trees – will dictate Australia's overall efforts to manage both water and carbon.

Planned fire is a crucial tool in the forest manager's toolbox. Without it, the risk of uncontrollable fire in much of Australia's southern eucalypt forests increases to become a certainty, and brings with it very serious risks to people and property. We have tried to write this book in a style that can be read with ease by anyone with a little basic scientific knowledge. It will hopefully be of some use in teaching at a wide range of levels, but we would like it also to be a stimulus for those interested in some of the more detailed scientific issues.

Finally, we hope this book will bring city and country people together through better knowledge. The increased risks to life and property associated with the shift out of the bush of families who lived and worked there for generations and who understood the need to manage fire with fire, has been compounded by the 'greenchangers' who move from the city to the country with little or no fire knowledge. Likewise, too many city people like to think 'the bush' will always be: (a) a haven for wildlife; (b) green; (c) an eternal sink for carbon and a source of water; (d) comforting in that it is out there, adding value to city lifestyles, views and values; and (e) not needing effort – 'nature will take care of itself'. Think again – for as long as people live on this land, we must manage it.

This book has not been written as a compendium of work on bushfires or fuel-reduction fires: that has been done already. Our intent is to bring together some of the key works in the theory and *practice* of bushfire management and fuel management. We have quoted liberally from some of the main researchers and practitioners – for example, Lachie McCaw and Rick Sneeuwjagt from the west, Phil Cheney and Kevin Tolhurst from the east, and David Bowman from the Northern Territory and Tasmania. We have also quoted liberally from Stephen Pyne (who has written eloquently about fire in Australia) and Jerry Williams (who has a wealth of experience in fire management) from the United States. In quoting the practitioners, we acknowledge that they may not always have tested their ideas or their data in the way scientists do, via peer-review. However, we believe it important to highlight their experience, for it is they who fight bushfires and to ignore their experience seems folly unless we are all willing to let bushfires burn. We have used a lot of our own published work; having written extensively about the subject ourselves, it is sometimes hard to re-cast our work in new words.

Finally, we write from our experiences – in forest management, in fire-fighting, and in forest research over many years. We have seen our long-term research sites severely burnt by bushfires in messmate forest at Mount Disappointment (November 1982), in mountain ash forest in the Britannia Range (February 1983, Ash Wednesday – Box 1), and in alpine ash forest in the north-eastern Divide (2007).

Box 1: Ash Wednesday, 16 February 1983 – a personal experience

Wednesday 16 February 1983 started off just like any other day in which politics dominates the news. A Federal election was due in just under 3 weeks, on 5 March, and the Prime Minister, Mr Malcolm Fraser, had given his policy speech the night before. Fraser had called the election on industrial law and order; the new leader of the Labor Party was Mr Bob Hawke, and so Fraser was 'fighting on Mr Hawke's traditional industrial relations battlefield'.[1] In his policy speech, Fraser said 'This is an election about responsibility, about the security that only responsibility can bring. Australia can't afford the turmoil and insecurity of the Federal Labor Party'.[2] Looking back, 27 years later, nothing seems to have changed much in election-speak, except for the players!

The forecast for 16 February 1983 was 'Cloudy. Mostly dry. Hot. Estimated top 34.' South-eastern Australia had been in the grip of one of the worst droughts on record. Rainfall for the preceding 12 months was less than 70% of average and, on 6 February, Melbourne was blanketed in a severe storm of red dust brought in from the north-west (Figure 1). Water restrictions were already in force throughout Melbourne, and new, tougher restrictions were to come in at midnight.

We left early on the morning of 16 February to travel up through the Yarra Valley to the Britannia Range, south of Warburton. Our mission was to inspect the fuel loads remaining after a clear-felling and harvesting operation in an area, or coupe, that was being logged by Mr J. Walker for Reid Bros., sawmilllers at Yarra Junction. We called first at Powelltown, to have a discussion with the District Forester (Yes! There was a Forests Commission and District Officers in those days!), and there the weather was warm, unearthly still and heavily overcast.

By the time we got up onto the Range, the sun had broken through and the temperature was soaring towards a maximum of 43°C. The logging track in from the forest road was thick with a fine dust. We had to close the windows, and the temperature within the car rose steadily.

Eventually we arrived at the log-landing, where logs were being loaded on a truck. Logging of this 22 hectare coupe was almost complete, with only a small

Figure 1.   The dramatic dust storm over Melbourne, Victoria, 8 February 1983.[3]

patch of mountain ash yet to be felled. Debris from the logging was piled high, in some places up to 2 metres. The routine process would be to burn this slash in the autumn, then to broadcast seed over the burnt area to ensure good regeneration. In fact, the coupe that was logged was a stand of mountain ash that regenerated after a major fire in 1926. But the ecology of mountain ash and its dependence on fire is a story for another chapter.

It was much too hot – If not much too dangerous – on that Wednesday to do much in the bush. We had a good look over the coupe, and came to the conclusion that trying to assess the mass and nutrient content of the debris lying on the ground was going to be a huge task. The wind was coming up from the north, and we took the wise decision to get out of the bush.

By 7 pm, the smell of smoke was obvious in the eastern suburbs of Melbourne, and charred particles were falling from the sky. A fire had been deliberately lit about 5.30 pm at the foot of Mt Little Joe, at Millgrove, and the Britannia Range was on fire. It turned out to be a major fire, burning more than 40 000 hectares of forested land over 6 days, 14 000 hectares of which was mountain ash forest.

Prophetically, Wednesday 16 February was the Holy-Day of Ash Wednesday. While the political battlefield had dominated the news, nature's battlefield had been steadily preparing for fire. The results for the Ash Wednesday bushfire near Warburton were devastating.

At noon on 15 February, the day before Ash Wednesday, 'a hot north-north-westerly airstream was moving over Victoria as a result of a ridge of high pressure off the east coast of Australia'.[4]

By noon on 16 February, this:

> 'high-pressure system had not moved and a deep trough of low pressure with several cold fronts had moved up over the Bight. As the pressure gradient increased the north-north-westerly winds also increased until by mid afternoon speeds of 70 km/h were recorded at Avalon (near Geelong) and Melbourne Airport.
>
> Temperatures were high and Melbourne Airport recorded 43°C at 1500 hours together with a relative humidity of 5%. Under these conditions the fuels were extremely dry. For example, fine fuels collected near Stawell at 1330 hours had a moisture content of 2.7% of oven dry weight.
>
> With the passage of a cold front across the State, a violent change of wind to south south-westerly occurred. This change reached Avalon at 2000 hours and Melbourne Airport at 2040 hours where maximum speeds of 105 km/h and 100 km/h respectively were recorded. Strong winds were maintained long after the front had passed'.[5]

These climatic forces were devastating for the Warburton fire. Even though it started in the late afternoon, the fire spread quickly (12 kilometres per hour) in a south-easterly direction toward Powelltown and then, following the violent change of wind with gusts up to 100 kilometres per hour, raced on a broad front in a north-easterly direction into the Upper Yarra catchment. The fire destroyed 23 houses, but no lives were lost.

We had been studying a number of aspects of the ecology of mountain ash forests at Britannia Creek for some 4 years before Ash Wednesday, 1983. It is a good place for scientific work because it includes some of the fastest-growing stands of mountain ash in Victoria. After the bushfire, we identified a range of sites in which the forest had burned at varying intensities – unburnt forest, light surface fire, crown fire and tree-kill and, the most intense of all, the logged coupe where all of the slash had burned ferociously. Over this range of sites, we have studied processes of recovery and regeneration of the magnificent mountain ash.

The Britannia Range burned again in 1991. A fire was lit at Millgrove near Mt Little Joe – more or less exactly the place where the 1983 fire started – on Thursday 7 March 1991. Only 8 years after Ash Wednesday, the local residents were terrified, as the local newspaper reported:

> 'Thursday was fearful. It was much too close. Thursday night we hoped the worst was over. Friday, fires flared and winds were temperamental. It was not over. There had been similarities with Ash Wednesday. There didn't seem to be the spread of fires but through that very hot Friday night into Saturday morning, there were real fears of an Ash Wednesday repeat. There was talk of a 90 km wind coming from the north and at 1 a.m. on Saturday we saw (from Yarra Junction) the horizon ablaze from Wesburn to Powelltown. Gladysdale was under threat and fire was racing towards Tarrango Road.

With a drop in the wind, some decent rainfall and lower temperatures by noon, Saturday, we think we can safely say that it is over now and, thank God, there were no human lives lost.'[6]

We were deeply worried. The 1991 fire came within a kilometre of our by-then well-established Britannia Creek Project. The investment in research and in the projects of a number of post-graduate students was close to being destroyed.

The drought of 1982–83, combined with weather conditions, was a dreadful fire season for Victoria. A major, El Niño drought developed strongly through 1982, and south-eastern Australia was tinder dry. In August 1982, 3400 hectares of the Little Desert burned; during November 1982 there were extensive fires through the Dividing Range including almost 21 000 hectares at Mt Disappointment not far to the north of Melbourne; we had long-term research plots in the Mt Disappointment forest, and they too were burnt. In December 1982 almost 18 000 hectares of the Big Desert burned. Two Forests Commission men were killed in a fire at Greendale on 8 January 1983, and fire started by a lightning strike burnt 127 200 hectares near Cann River in East Gippsland. On 1 February, 24 houses were destroyed by fire near Mt Macedon.

But worse was to come on Ash Wednesday, including a second fire at Mt Macedon. The Ash Wednesday fires burned in excess of 180 000 hectares of public and private land. Although the big fire at Warburton that we have described above was largely in forested areas, the fires at East Trentham-Macedon, Deans Marsh-Lorne, Upper Beaconsfield and Cockatoo raged through populated areas. The rate of spread of some of these fires was astonishing. For example, in the Deans Marsh-Lorne fire, the speed was 22 kilometres per hour over a 5-kilometre section, and the fire was spotting up to 10 kilometres ahead.

For the Ash Wednesday fires in Victoria, 'losses of life and property were of alarming proportions. In all, 47 people died, 2080 homes and 82 commercial properties were destroyed, 1238 farms damaged, 5900 km of fencing destroyed and some 7000 cattle and 18 000 sheep lost'.[7] Most of the deaths were in that dreadful hour immediately following the climactic wind change. Near Cockatoo in the Dandenong Ranges, 17 fire-fighters were trapped and died in their trucks with the unexpected force of the change. The personal tragedies of the Ash Wednesday fires were so great – both in Victoria and in South Australia where 28 people were killed by fire – that both the Liberal Party and the Labor Party were forced to cut back on their election advertising, but only for a couple of days and then it was back to business as usual.

Introduction

'Lives and property will be lost in future to bushfires and the most at risk will be at the urban–rural interface and in the tall-open forests'.

This short statement was part of a review of fire issues and research into those issues. The review was written by Professor Mark Adams at the request of the Minister for Science and was presented to the Bushfire Research Advisory Group in October 2002. That group paved the way for the current Bushfire Cooperative Research Centre.

There is nothing terrifically prescient about that statement – history says that the tall-open forests around Melbourne are some of the most fire-prone and dangerous in the world. Perhaps due to its matter-of-fact simplicity, it made little impact. If perhaps it had said: 'a rare and iconic species will be wiped out by fuel-reduction burning' it would have had more impact within the ranks of government officials in the offices of land management and fire agencies in Melbourne, Sydney, Perth, Adelaide and Hobart. Perhaps it should have said: 'Climate change will cause fire to destroy the forests', if the aim was to generate public support and gain a newspaper headline. Instead, it did little apart from perhaps help establish the need for a bushfire research centre. It certainly attracted only passing attention from the heads of the relevant agencies.

However, in the years since, beginning with the 2002 Sydney bushfires a couple of months later, the Victoria/New South Wales/Australian Capital Territory bushfires of 2003 that swept in to Canberra destroying more than 500 homes and killing four people, the 2006–07 bush-fires that raised the total area burnt in south-east Australia to more than three million hectares in just 5 years, and the worst civilian tragedy of the Black Saturday 2009 bushfires in Victoria in which 173 people lost their lives, bushfire has returned to the public conscience.

It should never have gone away. Most of all because of the major fires that still live in the memory of many and that are translated into prophetic words and recommendations by Royal Commissioners such as Judge Stretton. The fires of 1939 (Black Friday, Victoria), 1961 (Dwellingup, Western Australia), 1967 (Hobart, Tasmania) and 1983 (Ash Wednesday, Victoria and South Australia) scarred all involved. Not to mention the extensive fires in New South Wales including those close to Sydney in 1994. Yet go away our awareness did – more slowly for those affected and much more quickly for the great majority of the population living in our capital cities.

In part, concern about bushfires has been replaced by an urban concern about the use of planned fires or prescribed fires. An urban concern that speaks as much about the increased separation of city and country, of the increasingly poor knowledge of city people as to what is required to live in and manage the bush, as it does about the discomfort of smoke-tainted washing or wine, or conscience-twanging images on the nightly news or rants in major daily newspapers by columnists and others who make a living from the media, about 'escaped pre-scribed fires or backburns'.

Is it just coincidence that the recent unprecedented period of major bushfires follows on from one of the worst droughts on record in south-east Australia? Is it just coincidence that the rate of increase in the global concentration of carbon dioxide (CO_2) and in average annual temperatures is now greater than ever? Is it just coincidence that, since the mid-1980s, governments of all political persuasions in the eastern states have 'opted out' of fuel-reduction burning to control fuels (as opposed to the strong program of fuel reduction developed since the early 1960s in Western Australia)? These are all themes that we try to cover in this book.

In keeping with the above comments about the environmental knowledge of the general public, we make a point of addressing key issues about fires and ecological sustainability. Here we have adopted three of the globally recognised principles of ecological sustainability from the 1992 Australian Government's National Forest Policy Statement (Commonwealth of Australia 1992). They are that management should encompass and embrace:

- maintaining the ecological processes within forests (the formation of soil, energy flows, and the carbon, nutrient and water cycles)
- maintaining the biological diversity of forests
- optimising the full range of environmental, economic and social benefits to the community from all uses of forests within ecological constraints.

Many people in Australian capital cities and in the country are now only too well aware of the dollar value of water. Likewise, Australia and the rest of the world sits on the brink of using economic instruments such as carbon trading to try to combat the ever-increasing problem of rising atmospheric concentrations of CO_2 and their flow-on effects for temperature, rainfall and other aspects of climate. These are issues that transcend state and country borders. They are long-term issues that are crucially related to fire and must be tackled if Australia is to move toward ecological sustainability.

There are other similar issues – how does fire affect nutrient supply in the long term? How does fire affect soil formation processes? Will the energy balance of forests be affected by fire? Some of these are best exemplified by the frantic efforts to keep fire out of catchments owing to the massive pollution of waterways and dams that can be caused by thunderstorms and soil erosion after the fires have gone.

Purported negative effects on biodiversity are in large part a reason why the urban populace is concerned about planned fires, yet it remains curiously complacent about the obvious killing of wildlife and other effects on biodiversity caused by massive bushfires! Much of the general public's concern about planned fires has been fuelled by scientists seemingly unable or unwilling to see a larger picture. We try here to set the record straight.

We have drawn from a wide range of sources, including relevant overseas experience and knowledge, particularly where we discuss issues such as water, carbon, nutrients and soil formation. Science is a universal language and molecules of carbon, water and nutrients behave according to the laws of physics and chemistry as well at home as they do abroad. Evolution too is now a universally understood and accepted concept in biology, and the biota of Australia's southern forests has evolved with fire and because of fire. The best use of the so-called precautionary principle is then to recognise that withdrawing fire from forests will have major deleterious consequences for the biota.

In summary of the above, the content of this book can be divided into our efforts to provide answers to two questions: (1) To what extent can we protect life and property from bushfires in our southern forests? (2) What role does management have in ecologically sustainable fire regimes in our southern forests?

The rest of this chapter provides introductions to these questions.

Lives and assets at risk

A useful first step is to examine what is at risk and why it is at risk.

Most at risk are those people living on the fringe of our cities, either by choice or necessity. The increase in population in the 'fringe' areas of major metropolitan centres is probably inexorable. As described in an email to one of us from Stephen Pyne, the noted fire historian:

> '*The developing world is witnessing a migration to cities; the developed world, a parallel migration out of cities. The intermix is a phenomenon of the developed world as an urban out-migration reclaims a previously rural landscape. Hence, we find urban values, urban expectations, urban aesthetics, an urban economy – and a falling away from rural bush practices. My sense is that this is a rather large tide, not easily stemmed, save locally.*'

Let's consider the Upper Yarra Valley and Dandenong Ranges (UYVDR) area to the east of Melbourne. In this region of some 280 000 hectares, there remains about 230 000 hectares of forested land. Over the 50 years since the end of the Second World War, the area of forested land has changed little. Public land accounts for approximately 73% of the UYVDR and includes most of the water supply catchments and forested areas close to Melbourne. Private land used for rural activities comprises approximately 24% of the region's total area and urban development accounts for 3% of the total area but a high proportion of the population.

The Upper Yarra Valley and Dandenong Ranges rank among the most fire-prone areas in the world. Major fires swept the area in 1851, 1898, 1926, 1932, 1939, 1962 and 1983. The bushfires of 14 February 1926 killed seven people and large numbers of livestock. A considerable portion of the region was burnt during 1932, but those fires were minor by comparison with the 'Black Friday' fires on the 13 January 1939. Those fires claimed 71 lives and killed vast numbers of trees in the UYVDR. Similar major fires in 1983 were again of stand-replacing intensity, burning through almost 40 000 hectares of prime mountain ash (*Eucalyptus regnans* F. Muell.) forest in the UVYDR and claiming many lives including, tragically, those of some fire-fighters.

Over the same 50-year period, the population of the Upper Yarra Valley and Dandenong Ranges has increased five-fold, with similar increases in the number of dwellings and a concomitant decrease in the number of rural holdings (Table 1.1). Land use has changed dramatically from low-intensity grazing (sheep) to high-value agriculture, viticulture and horticulture.

The rising value of the available private land, driven by its potential availability for housing, underpins and drives changes in land use and planning strategies. Increases in production value (e.g. replacement of wool production by winemaking) help ensure land remains at least partially 'rural', instead of residential. Nevertheless, both increased numbers of houses and increased intensity of use of rural land, increase the infrastructure at risk.

The net value of property in the Upper Yarra Valley and Dandenong Ranges is now many-fold greater than it was in 1983, let alone 1939 or before. Equally, a high proportion of residents are now 'recent arrivals', with little knowledge either of previous fires and their behaviour (including rapid rates of spread) or the flammability of their environment. The lack of fire knowledge of the increasingly 'urbanite' population is compounded by the loss of 'bush' people (e.g. the decline in rural holdings, Table 1.1) – often extended families and small communities who were accustomed to being self-reliant in the face of the frequent fires.

Although the above is a summary of land-use changes on the fringe of Melbourne, it could have been prepared for any of the Adelaide Hills, the hill suburbs to the east of Perth, or the

Table 1.1.   Historical changes in land use within the Upper Yarra Valley and Dandenong Ranges region (after Kasel 1999).

Year	1947	1967	1993
Population	29 707	55 530	148 927
Year	1966	1976	1991
Number of dwellings	18 588	30 680	49 583
Year	1959	1977	1993
Number of rural holdings	1981	965	477
Total area of rural holdings (hectare)	81 745	49 530	26 033
Year	1959	1977	1994
Number of beef cattle	12 682	39 905	25 384
Number of pigs	5865	11 527	13 395
Number of dairy cattle	18 771	9 796	2530
Number of sheep	50 725	10 067	1506
Year	1977	1990	2003
Grape production (hectares)	75	415	>2500

dissected sandstone terrain of outer Sydney. Some locales are more topographically complex than others; some have water catchments and some not; some have more urban development and less intensive agricultural development. Irrespective, the core issues remain – increased numbers of people and properties on the periphery of capital cities. Others have noted that this expansion consumes a great deal of public funding. The provision of services such as roads, water, power, communications and sewerage, and of the accompanying needs of communities such as schools, shops and health care, becomes increasingly costly as the distance from the centre of the metropolis rises. This cost is largely borne by governments. Fire and the costs thereof, is another reason why governments might consider acting to reduce the sprawl.

Beyond the urban–rural interface, there are many other lives and much property at risk. Of greater spatial extent, and therefore potential cost, has been the risk posed by the loss of resources – experienced people and infrastructure such as tracks and machinery – from the large areas of native forest in south-eastern and south-western Australia. John Dargavel (1994) described how socio-political and economic forces have shaped the nature of forested landscapes. The rise of environmental groups and their success in convincing the urban populace that logging of native forests was equivalent to land-clearing (however incorrect) led to withdrawal of government support and the subsequent withdrawal of private capital. In 2008, we described much of the change in tenure of forested land, from state forest to national park, as well as the industry and conservation outcomes (Attiwill and Adams 2008).

In 2008, we also described how states and the Commonwealth have been active in promoting industrial plantations, mainly for the purpose of producing woodchips for pulp and fibre markets. Government initiatives have been a part of the expansion of the plantation estate and the increased supply of wood products from plantations. As an example, more than 250 000 hectares of eucalypt plantation have been established in south-western Australia alone in the past 20 years (Colour plate 1) and Australia-wide the area is several-fold greater. Some of these plantations are contiguous with native forest and many formally isolated (within agricultural land) plantations have coalesced into large blocks. The development of the largely privately owned plantation estate has not always been accompanied by an increase in resources for fire management, or if there has been such investment, it is often fairly recent.

The nature of capital investment in forestry operations, both in native forests and plantations, has also changed, with special-purpose harvesting and haulage machinery replacing the previously generic bulldozers. The newer machinery is less adaptable to fire-fighting purposes. Returning to Dargavel's view, the rate of movement of 'capital' is perhaps faster than might have been anticipated in 1994. Dargavel's concluding remarks were that the influences of the environmental movement and industry ('development at home and markets abroad') would shape the forested landscape in future. However, it is fire, far more than logging, that has shaped the forests ever since.

Fire risks are increasing as a result of: (1) the increased area of plantations developed for local and external markets; (2) withdrawal of private capital from state forests and the gradual process of redirection of state resources; and (3) increased capital investment and number of people at the urban–rural interface.

Fire regimes and risks to ecological processes and biodiversity

The very large number of scientific publications on fire in Australia is dominated by studies of the effects of fire on biodiversity. The collation by Bradstock *et al.* (2002) on fires and biodiversity was complemented a year later by Abbott and Burrows (2003), with an equally comprehensive synthesis of fire knowledge of south-west Australia. The 38 chapters by dozens of authors in these two works alone cite more than two thousand primary references. When coupled with the many other recent works on more specific communities and ecosystems by, for example, Andersen *et al.* (2003), Bowman (2000) and Mackey *et al.* (2002), it is small wonder that Stephen Pyne noted (2003): '*The amount of research is astonishing*'.

The large amount of research in the south-west of Western Australia has allowed the formulation of a number of clear scientific principles to guide forest management (Burrows and Abbott 2003). However, this is not so for the rest of southern Australia and, like much of biology, it is a complex story that needs a good deal of interpretation.

In part, we lack simple rules because we have been recording the biological responses of plants and animals to fire for a relatively short period. Not much of the specific literature dates back more than 30–40 years. What we have learnt is largely the response of organisms to single fire events. Understandably, scientists then point out that it is not the single fire event that matters. It is the overall fire regime – how many fires over a given period and of what intensity – that matters and then we lack all but the most rudimentary direct evidence because there are literally no studies that have encompassed a time scale long enough to be meaningful. There are no rigorous studies that have included the combined effects of planned and unplanned fires. We also face problems when attempting to define the pre-European fire regime over much of southern Australia (see Box 1.1). There is some empirical evidence, but much more conjecture. While a pre-European 'state' might be the preference for some people based on aesthetic or philosophical grounds, it is inconceivable to us that current Australian society could or should aim to 'turn back the clock'.

In the absence of hard data on the effects of *fire regimes* on biodiversity, scientists interpret what might be the effects of fire regimes based on their observations from many sites and separate studies. Sometimes they have used what are called 'space-for-time' comparisons or chronosequences. Perhaps these are best imagined as a set of study areas, all within the same forest, but where each area has had a different period of time elapse since the last fire. These

Box 1.1:   Aboriginal and European fire regimes?

Pyne's analysis of Australia's fire history (1991) and similar works by other historians such as Flannery (1994) have maintained the debate about pre-European fire regimes. The debate has continued in some quarters because Aboriginal Australians were widely regarded as skilled users of fire, and in particular knew how to use 'cooler or at least more manageable fires' to achieve their objectives. This has become a touchstone issue for those like to argue about the 'ecological significance' of European use of planned fires.

Some geographers, such as Scott Mooney at the University of New South Wales, have argued that Europeans have increased fire frequency (e.g. Mooney et al. 2001). Much of the evidence they use is point-based – tree rings and soil/sediment cores are the most common sources of information with charcoal a key diagnostic tool. Some of this evidence has accumulated from larger areas such as catchments, albeit with little or no information as to how much of a catchment was burnt or at what intensity. However, much of it remains evidence of local, not regional, fires. Even then, it may not capture fires of lesser intensity (e.g. Bowman 1998).

Others point to the many written and verbal accounts of fire frequency at the time of European settlement and in the hundred or so years that followed. These accounts, often written by early European explorers (see Enright and Thomas 2008 for a summary) suggest that Aboriginal Australians used fire widely and frequently in southern Australia, not just in the north. Nevertheless, some types of vegetation (e.g. dense wet forests and shrublands) seem unlikely to have been the focus of Aboriginal burning. On balance, Enright and Thomas concluded that the 'fragmentary' evidence for the south provides little clarity. Dry sclerophyll forests were probably deliberately burnt more often than the wet forests. Some Aboriginal fires probably got out of hand to become bushfires.

While the European/Aboriginal debate will continue, it will become increasingly irrelevant as time passes. Australia today is not the Australia of pre-European settlement. There has been more than 200 years of continuous change to fire regimes and that is not about to change – the rate of change will more likely increase. It makes more sense to conclude, as did Enright and Thomas (2008) that:

> 'any increase in the area burned, and decrease in fire interval associated with such a strategy, will need to remain based on scientific criteria that seek to meet current (and evolving) biological conservation and asset protection objectives rather than being based in any sense on uncertain past regime.'

We would modify that statement such that the criteria should meet objectives for ecological sustainability – including, for example, carbon, water and nutrients – as well as asset protection.

types of studies are useful, but far from definitive – there are too many confounding influences. Consequently, much of the literature ends with statements about 'potential threats' to biodiversity from inappropriate fire regimes.

Hence, in place of definitive evidence as to the effects of fire regimes on this or that plant or animal species, we have computer models and much conjecture. There is no doubt that land managers would like to be able to predict the effect of fire regimes on biodiversity, but the current models are little more than guesswork, relying as much on ecological theories such as succession, as on confirmed mechanisms or empirical data. That is unlikely to change for many years. The models, many of which are laden with challengeable assumptions, may provide some guidance, but their confirmed ability to predict the numbers or identity of species, or their abundance in time and space, is so negligible as to be of little practical value – later in this book we tackle this issue again in some detail.

By way of introduction to the other elements of ecological sustainability and the effects of fires thereon, it is useful to have an appreciation of forests generally. For example, the Centre for International Forest Research in Bogor, Indonesia, recently (2009) highlighted the fact that the world's forests account for more carbon emissions than do cars, trucks, airplanes and ships. These emissions are dominated by fire and especially by the effects of fire on forest soils that hold vast amounts of carbon. Likewise for water – the world's great forests provide the water for the world great rivers. It is no different in Australia. The forests of the Great Divide currently provide the water for the cities of Adelaide, Melbourne, Canberra and Sydney, as well as the water used by agriculture in Australia's food bowl – the Murray–Darling Basin. In the west, the jarrah forests of the Darling escarpment provide much of Perth's water. Analysis of the effects of fuel-reduction fires on carbon, water and a host of other ecological attributes, in addition to biodiversity, seems essential to us.

2

Bushfires in Australia

An overall view

Fires before European settlement

Australia, in its northward drift from the Gondwanan continent over the past 150 million years or so, has now reached the position where it is widest at the driest part of the world – the Tropic of Capricorn. During this northward drift, the vegetation over the continent changed from one dominated by Gondwanic cool-temperate rainforest (*Nothofagus* spp. that we see today only in the most sheltered and wet gullies of south-eastern Australia) to one dominated by *Eucalyptus* and other genera with hard, 'sclerophyllous' leaves and with a high degree of endemism (native to an area). With increasing aridity and sclerophylly came an increasing incidence and spread of fire from lightning. At some time in the late Pleistocene – estimates vary within the range 45 000 to 70 000 years ago – Aborigines came to Australia. The frequency of fire increased dramatically:

> '*Within Aboriginal society fire was pervasive. It assured Aboriginal dominance over megafaunal Australia much as it had confirmed eucalypt dominance over megafloral Australia. Fire and Aborigine were never far apart; on this the hottest and driest of the vegetated continents, Aborigines habitually walked across the landscape armed with smouldering firesticks*' (Pyne, 1991, p. 84).

> '*In the Aborigine, Australian fire had discovered an extraordinary ally. Not only did ignition sources multiply and spread, but fire itself persisted through wet season and dry, across grassland and forest, in desert and on mountain. Lightning was a highly seasonal, episodic ignition source; the Aboriginal firestick was an eternal flame. The domain of fire expanded, not only geographically but temporally, for this inextinguishable spark obliterated even the seasons. But 'if fire was maintained by the Aborigines, it is also true, as Phyllis Nicholson (see Nicholson, 1981) notes, "that the Aborigines were maintained by fire." The relationship between them was reciprocal, symbiotic. "The evidence that fire was the indispensable agent by which Aboriginal man extracted many of his resources from the environment is irrefutable"*' (Pyne, 1991, p. 85).

The early explorers (dating back to Abel Janz Tasman, December 1642) and first settlers provided many, quite detailed records of the nature of the Australian bush and of extensive fires. A recurring theme is the open, grassy nature of forests and woodlands, compared with the understoreys of woody shrubs that we see today. Abbott (2003) and Ryan *et al.* (undated) published collections of early records, together with interpretations and syntheses by modern historians. The Ryan *et al.* collection begins with a quote from Tim Flannery, *The Future Eaters* (1994):

> 'As a result of the extensive Aboriginal use of fire, the plant communities seen and recorded by Banks and other explorers were very different from those that exist at the same locations today ... The lack of recognition of this change has critical implications for the management of national parks and wilderness areas today'.

The collection concludes with this quote from Geoffrey Blainey's *Triumph of the Nomads* (1982):

> 'Thousands of years of burning could not fail to affect the landscape and all that lived on it. The sheep-owners who came from Britain did not have the faintest idea of how long the Aboriginals had occupied the land but they had a sound idea of the botanical effects that came within a few years of the cessation of burning. If five or ten years that experienced few fires could alter the vegetation of Australian forests and grasslands, it would not be surprising if thousands of years of fires had also altered the previous vegetation ...
>
> Without those fires the grassy woodlands that occupied much of the fertile crescent in south-eastern Australia would have been scrubland or forest. A period of fifty years was probably sufficient to change the character of that savannah country if no fires burned ...
>
> Fire was also an emblem of the collapse of their (Aboriginal) society. By helping to create many of the grasslands of the south-east, fire indirectly attracted the Europeans and their sheep and cattle to the interior and so quickly extinguished a way of life which was essentially pastoral.'

David Bowman (2003a) provides an excellent but simple summary of the changing forces of fire in the evolution of the Australian biota (Figure 2.1). Over hundreds of thousands of years, lightning started fires that created a broad scale mosaic (Figure 2.1a). Aboriginal management of the land obliterated this broad-scale mosaic, and created a much finer-scale mosaic (Figure 2.1b). While acknowledging the contentious debate over the causes of extinction of the Australian megafauna, Bowman states that the Aboriginal fire regime may have at least been a contributing factor – 'that fire may have sufficiently changed the understorey characteristics of many woody vegetation types, including tropical savanna, as to have ultimately disadvantaged the megafauna.'

In turn, European fire regimes completely obliterated the fine-scale mosaic of Aboriginal burning (Figure 2.1c), destroying the diversity of habitats and creating high fuel loads over expensive areas; the result is 'feral' fire at a coarser scale then ever before.

Bushfires in the modern era

Fires burnt some 3 million hectares in Victoria and New South Wales over the summer of 2002–2003. In the calendar years 2002–2003, fires burnt 95.4 million hectares throughout Australia (12.4% of the total area), about 80% of it in the tropical savannas to the north, as we can see by the distribution of fires across the continent for 2002–2003 (Figure 2.2).

For the southern forests that we are concerned with in this book, the average annual (1995–2004) areas of forest burnt by bushfires and by fuel reduction are shown in Figure 2.3. It is rather difficult to analyse these data; the bushfire data are for all types of forest, from the wettest eucalypt forests to the driest woodlands, while much of the fuel-reduction burning is

Time period
since last fire

☐ Long
▨ Medium
■ Short

(a) Pre-human Fire Regime   (b) Aboriginal Fire Regime   (c) European Fire Regime

Figure 2.1.   In the pre-human period (a), lightning started fires infrequently and burnt large areas, creating a broad-scale habitat mosaic to which various species of birds and mammals had become became adapted. Aboriginal fire management (b) was characterised by a high frequency of fires that burnt much smaller areas, producing a fine-scale habitat mosaic that supported most of the pre-human wildlife assemblage, with the notable exception of the Pleistocene megafauna. Under European fire management (c), fires, that had a similar frequency as the Aboriginal period, burnt large areas thereby obliterating the pre-existing habitat mosaic created by Aboriginal landscape burning. (Figure and caption from Bowman 2003a.)

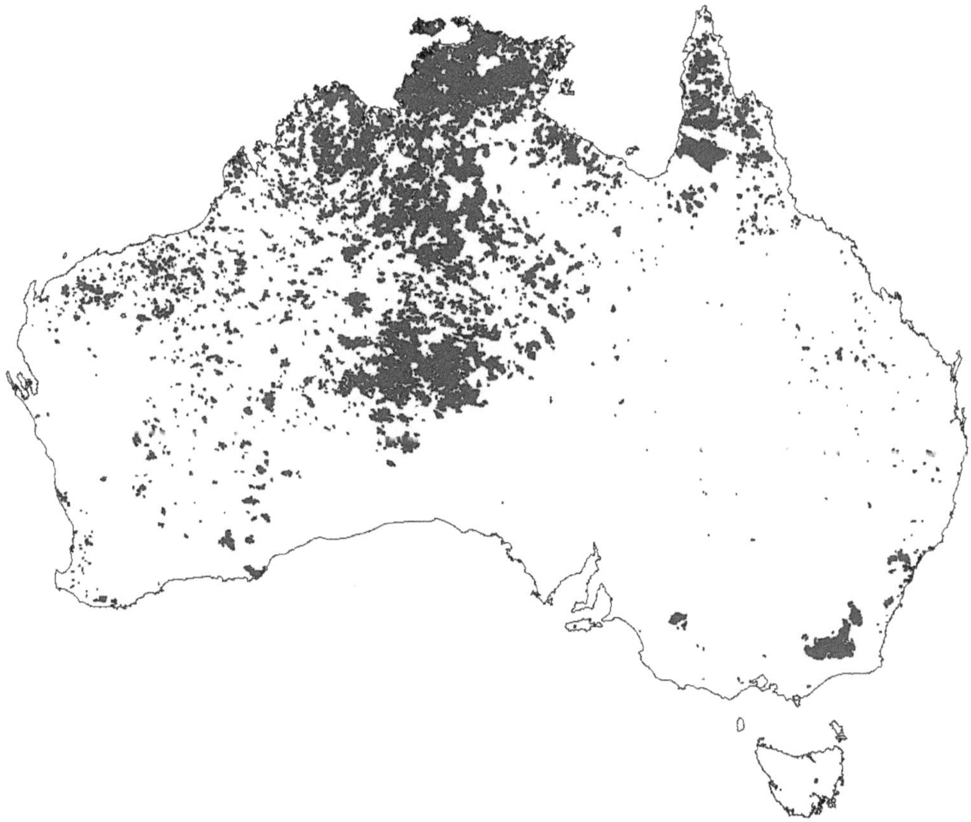

Figure 2.2.   Fires in Australia, April 2002 to April 2003 (courtesy Belinda Heath, Landgate – West Australian Land Information Authority).

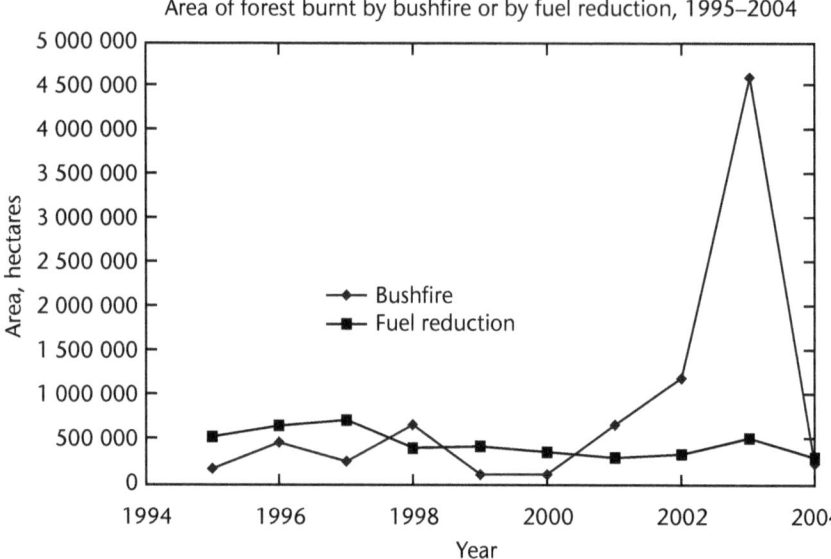

Figure 2.3.   Areas of all types of forest burnt by bushfire in the southern areas of Australia (the forests of Western Australia, South Australia, Tasmania, Victoria, Australian Capital Territory and New South Wales). Data from Department of Environment, Water, Heritage and the Arts.

restricted to specific areas (for example, the forests of the south-west of Western Australia). However, it is clear that the annual area of fuel-reduction burning was relatively constant, and that 2002–2003 was a bad bushfire year. Over the 10 years 1994–2004, the ratio of area of forest burnt by bushfire to the area of fuel-reduction burning was lowest in Western Australia (at 1.4), between 2 and 2.6 for New South Wales, Victoria and Tasmania, and an astonishing 188 for the Australian Capital Territory where a meagre 0.08% of the forest estate of 120 000 hectares is burnt for fuel reduction annually.

Bushfires in the southern forests

Major fires following European settlement
The following summary comes from a number of sources, principally Clack (2003), Department of Sustainability and Environment Victoria (2010b), Dexter (2005), Romsey Australia (2010) and State Government of Victoria (2003). There are some discrepancies between the various records; for example, the Department of Environment, Water, Heritage and the Arts states that 'over 67 million hectares of Australia burnt in 2002–03 due to unplanned grass and forest fires' with the acknowledgment that the 'extent of bushfires in 2002–03 has not been verified. There are large variations in estimates given'. We have tried to be as accurate as possible; however, we acknowledge that errors are inevitable, given the uncertainty of some of the historical records.

Early 1700s
Victoria – Before the fires of 1939, forests of mountain ash estimated to be about 200–250 years old covered much of the Dividing Range. The fire-dependent regeneration of these forests ensures they could only have originated after a widespread fire of stand-replacing intensity in

the early 1700s. The 1939 fire killed most of these old forests, so that the mountain ash forests of the Dividing Range are now just over 80 years old, with only a few stands of old-growth forest, about 350 or so years old, remaining.

1832

Tasmania – Extensive fires surrounding Hobart. 'On several evenings during the week Hobart-Town has been illuminated by brilliant but awful flames on the sides of the hills, overhanging the town to the imminent danger of the neighbourhood. And on the other side of the Derwent as far as the eye can reach the horizon has been one range of fire and smoke.'

1851

Victoria – Black Thursday, 6 February – one-quarter of Victoria burned, fires 'from Gippsland to the Murray, and from the Plenty to the Glenelg'. The Dandenong Ranges were devastated. Temperatures at Melbourne reached 112–117°F (44–47°C). About 12 people killed, 1 million sheep and thousands of cattle destroyed, and some 5 million hectares burnt (a quarter of Victoria).

1854

Tasmania – 1 January – Fires along the Huon River at Cygnet and Esperance 14 people reported to have lost their lives.

1897

Tasmania – 31 December – fire starting at Mt Wellington killed six people, and burnt for a week, threatening Hobart.

1898

Victoria – Red Tuesday, 1 February – Fires in South Gippsland burnt 260 000 hectares, killing 12 people and destroying 2000 buildings.

1926

Victoria – Black Sunday – 60 people died, including 31 people over the weekend of 13–14 February. The fires were spread around Noojee, Kinglake, Warburton, Erica and the Dandenong Ranges.
    New South Wales – 1926–27 season – more than 2 million hectares and eight people killed.

1932

Victoria – many fires, particularly in Gippsland. Nine people killed.

1939

Victoria – Black Friday, 13 January – total area burnt between 1.5 and 2 million hectares, 71 lives lost. Large areas of the north-east, Gippsland, the Otway Ranges and the Grampian Ranges affected. The town of Narbethong destroyed. More than 650 homes or shops destroyed, 69 timber mills destroyed.
    New South Wales – six people killed in southern and eastern New South Wales.

1942

Victoria – 3–4 March – Fires in western Victoria and South Gippsland. One person killed, 200 sheep, two farms and more than 20 homes destroyed.

1943

Victoria – Fires at Wangaratta – 10 people killed, and thousands of hectares of grass country burnt.

1944

Victoria – January to February – Central and western Victoria, Yallourn. More than 1 million hectares of grassland and 160 000 hectares of forest burnt. Forty-nine people killed, 500 homes destroyed and huge losses of stock.

1952

Victoria – 100 000 hectares burnt in north-east Victoria, several lives lost.

1955

South Australia – Black Sunday, 2 January – Adelaide Hills.

1957

New South Wales – Blue Mountains – fire driven by gale-force winds destroyed 25 homes, shops, schools, churches and a hospital.

1958

South Australia – severe fire in a pine forest at Wandilo; eight fire-fighters killed.

1960–61

Western Australia – 33 major fires raced across 350 000 hectares in the north-west and central pastoral regions, 1.5 million hectares burnt in the southern pastoral region. The Dwellingup fire burnt 145 000 hectares of jarrah forest.

1962

Victoria – 14 January – fires on the north-eastern and eastern outskirts of Melbourne (The Basin, Dandenong Ranges; Kinglake, St Andrews, Hurstbridge, Warrandyte, Mitcham) – 32 people killed and 454 homes destroyed.

1964–1965

New South Wales – major fires in the Snowy Mountains, Southern Tablelands and Sydney's outer metropolitan area. The Chatsbury-Bungonia fire burnt 250 000 hectares, and the Tumut Valley fire burnt 80 000 hectares.

Victoria – January to March – fires at Longwood and in Gippsland. Seven people (all from one family) killed in fires over 15 000 hectares of grassland and 300 000 hectares of forest. More than 66 homes or shops, 4000 stock destroyed.

1967

Tasmania – Black Tuesday, 7 February – 110 fires burned, of which 20 accounted for 58% of the total burned area. Fire near Hobart incinerated suburbs as it went. More than 270 000 hectares burnt, 1300 houses and 128 major buildings destroyed, 62 people killed. The fires reached 2 kilometres from central Hobart, fanned by winds averaging 80 kilometres per hour.

1968

Victoria – Fire in the Dandenong Ranges. The fire was contained to 1920 hectares, but 64 homes and other buildings were destroyed.

New South Wales – 1 November – Blue Mountains – 14 lives lost, 150 buildings destroyed. Over 1 million hectares burnt in a fire lasting for 4 weeks.

1968–1969

New South Wales – widespread fire over much of eastern New South Wales. Fires at Wollongong destroy 33 homes. Blue Mountains, 123 buildings destroyed. Fourteen people killed and more than 1 million hectares burnt.

1969

Victoria – 8 January – Two hundred and eighty fires broke out, including Lara, Daylesford and Yea; 12 fires reached major proportions. There were 23 deaths (including 17 motorists in a severe grass fire over the Melbourne-Geelong Road near Lara) and more than 250 000 hectares burnt, 240 buildings destroyed and more than 12 000 stock killed.

1972–1973

Victoria – Mt Buffalo – 12 140 hectares burnt.

New South Wales – 200 000 hectares burnt in the south-east.

1974–1975

New South Wales – dreadful bushfire season in the far west. Some 3 755 000 hectares burnt, 50 000 stock lost, 10 170 kilometres of fencing destroyed.

1976–1977

New South Wales – early December – Blue Mountains, 65 000 hectares burnt.

Victoria – 12 February – Fires over 103 000 hectares of central and western Victoria. Four people killed, and more than 100 houses or shops and 200 000 stock destroyed. Spread rates recorded up to 16–17 kilometres per hour.

1977–1978

New South Wales – Blue Mountains – December 16 – about 100 000 hectares, 86 buildings and vehicles destroyed – but Clack says '49 buildings destroyed, and 54 000 hectares burnt'.

Victoria and South Australia – grassfires swept through Victoria to the South Australian border, killing five people, destroying 116 houses and killing almost 200 000 stock.

Western Australia – fires over 31 500 hectares from Wanneroo to Pemberton, one person killed.

1978–1979

New South Wales – Fires in the Southern Highlands and south-west slope regions. More than 50 000 hectares burnt, five houses destroyed, and heavy stock losses.

Victoria – Fires near Bairnsdale – two people killed and one home and 6500 stock destroyed.

1979–1980

New South Wales – some 1 million hectares burnt in total over most of the state.

1980–1981

Victoria – December–January – Fires in the Sunset country and Big Desert burn over 119 000 hectares.

1982–1983

New South Wales – $12 million worth of pine plantations destroyed in southern New South Wales.

1983

Victoria – 31 January – fires at Cann River, East Gippsland burned 250 000 hectares, most of it state forest.

Victoria – 1 February – fire at Mt Macedon burned 1860 hectares and destroyed 50 houses.

Victoria – Ash Wednesday, 16 February – Fires widespread, including Cockatoo, East Trentham, Mt Macedon, Otway Ranges, Warburton, Upper Beaconsfield. Forty-seven people died, 27 000 stock killed, and more than 2000 homes or shops destroyed. Around 210 000 hectares burnt.

South Australia – First 'Ash Wednesday' fire, 1 February – 51 homes destroyed in the Adelaide Hills. Ash Wednesday, 16 February – The South Australia fires burnt 208 000 hectares in the Adelaide Hills and in farming country in the south-east of the state. In addition, 21 000 hectares of pine plantations were destroyed. Twenty-eight lives and 383 houses were lost.

New South Wales – more than 1 million hectares of grassed land in the west was burnt. During the season, 6000 fires started, a total of 3.5 million hectares was burnt, four lives were lost and, 40 000 stock were killed.

1985

Victoria – 14 January – Avoca/Maryborough, Little River, Springfield, Melton – three people killed, 182 homes, 400 farms and 46 000 stock destroyed by fires over 50 800 hectares.

1987–1988

New South Wales – more than 115 000 hectares burnt in the Bethungra and Warurillah-Yanco fires, three lives lost. In south-eastern part of Kosciuszko National Park 65 000 hectares burnt.

1990

Victoria – 27 December. One person killed, and 17 homes and more than 12 000 stock destroyed.

1990–1991

New South Wales – November – Hay-Murrumbidgee – 200 000 hectares of grazing land burnt, 100 000 sheep killed, hundreds of kilometres of fencing destroyed. A week later, a further 80 000 hectares burned, 76 000 sheep and 200 cattle destroyed. December 23 – hundreds of fires across the state, eight homes destroyed.

1991–1992

New South Wales – October 16 – More than 30 blazes around the state. Two lives lost and 14 homes destroyed.

1993–1994

New South Wales – late December – January – More than 800 fires started, from the Queensland border to Bateman's Bay. More than 800 000 hectares burnt, four lives were lost and 206 homes destroyed. A huge fire-fighting effort with fire-fighters brought in from across Australia and New Zealand.

1994–1995

New South Wales – Sydney and eastern New South Wales – two fire-fighters and two others killed, more than 24 000 evacuated from their homes.

Victoria – 23 February – 10 000 hectares of forest burnt at Berringa.

1997

New South Wales – December – bushfires across the State and around Sydney. Two fire-fighters killed near Lithgow.

Victoria – January – three people killed sheltering in a house in the Dandenong Ranges. More than 40 homes destroyed. Fires in central Victoria (including Creswick and Heathcote) burnt 400 hectares.

Western Australia – Perth and south-western region. Two people died and 21 injured in three major fires.

1997–1998

Victoria – December-January – Alpine National Park, Caledonia River area – 32 000 hectares burnt.

1998

Victoria – five fire-fighters from CFA, Geelong killed in a fire of 780 hectares near Linton, south-west of Ballarat.

2000

Victoria – 17–19 December – 29 000 hectares of grassland burnt at Dadswell's Bridge, Western Victoria.

2001

New South Wales – Black Christmas Fire – 170 homes destroyed and hundreds of thousands of hectares of bushland burnt.

2002

New South Wales – Nearly 50 homes destroyed in Sydney's outskirts.

Victoria – December – Fire in the Big Desert burnt 181 400 hectares.

2002–2003

Victoria – Alpine fire – 8 January to 9 March – no fire deaths although one fire-fighter was indirectly killed, 1 067 500 hectares burnt. Forty-one homes and 9000 livestock destroyed. Kilometres of fencing destroyed.

Victoria and New South Wales – The Alpine fires – a total of more than 3 million hectares burnt.

Australian Capital Territory – the Canberra firestorm – Four people killed and more than 500 homes destroyed. The Observatory at Mt Stromlo destroyed.

Western Australia – The 2002–03 bushfire season in the southern half of Western Australia was the most severe in 42 years as a result of at least 5 years of drought and severe lightning storms. A total of 656 wildfires burnt 2.11 million hectares of land. With only two exceptions, all these lightning-caused fires were contained to relatively small sizes. In the south-west of the state, a total of 549 bushfires covered about 126 000 hectares of lands managed by Department of Environment and Conservation.

2005
South Australia – Eyre Peninsula, 145 000 hectares burnt, nine people killed, eight of them fleeing from the fire in their cars.

2006
Victoria – 160 000 hectares, Grampians National Park (50% of the entire Park burnt) and Anakie, Victoria, 2005–06 (60% on public land). Two people died in their car.

2006–2007
Victoria – 1.2 million hectares, Great Divide fires, 1 million hectares of public land burnt.
   Tasmania – fire widespread, particularly along the east coast.
   Western Australia – 3 February 2007 – 13 000 hectares burnt in and around Dwellingup. One person killed, fleeing from the fire. 28 December – 29 000 hectares burnt in Boorabbin National Park, one person killed.

2009
Victoria – 430 000 hectares, Black Saturday fires, 7 February (some 70 national parks and other reserves, including nearly all of the Wallaby Creek and O'Shannassy catchment stands of old mountain ash), 173 people killed.
   Western Australia – Toodyay fires 29–30 December, 37 houses destroyed and 3000 hectares burnt.

3

The nature of fire

The fundamentals of bushfire are not difficult to understand. A fire starts when there is a source of ignition and a supply of fuel. What happens thereafter is variable and complex, depending on the moisture conditions, nature, quantity and distribution of the fuel, on topography, and on the weather. We will not discuss the spread and behaviour of a bushfire in detail. That has been comprehensively outlined (Luke and McArthur 1978) by the fire champions of Australia's CSIRO – Harry Luke, Alan McArthur and Phil Cheney, and it is essential knowledge for those who have to fight fires. The basics and terminology of bushfires are also treated in detail by Kevin Tolhurst and Phil Cheney (1999). We aim here to give no more than a brief summary of fire and fire behaviour.

Sources of ignition

The major cause of fire over geological time in Australia is lightning. The frequency of fire increased with the arrival of Aborigines tens of thousands of years ago and increased again following European occupation a couple of centuries ago. But a quick glance at the world distribution of lightning strikes shows that lightning is regular and frequent over much of the world, and David Bowman (2005) suggests that there is a global association of high lightning activity with fire-adapted floras.

Lightning activity varies from less than 1 flash per square kilometre per year over much of Tasmania, about 4 flashes per square kilometre per year over much of the southern mainland, 10 flashes per square kilometre per year over much of the central regions, to 20–30 flashes per square kilometre per year in parts of northern Australia.

Even the relatively low frequency of lightning in southern Australia means that lightning is the major cause of fires in Victoria, just ahead of deliberately lit fires: the two together igniting 50% of all fires (Table 3.1). However, lightning strikes alone account for almost 50% of the total area burnt each year.

What can be more a part of nature than a fire caused by a lightning strike? Should we extinguish *all* fires started by lightning, or should we let some of them burn? That is a question that is asked around the world. However, we note here that Paul Collins (2006) has written a book strongly presenting the case against the use of fire in management. It is rather extraordinary that we can find only one reference to lightning throughout Collins' book (there may be more, but lightning – the key to much of the evolution of Australia's flora and fauna, and the major cause of ignition of fires today – does not even warrant an entry in the index!).

Fuels in the forest

Eucalypt forests are evergreen, shedding only 30–50% of their leaves each year. This fall of litter is mostly concentrated in the summer months. The annual fall of litter ranges over an

Table 3.1. Causes of fire on public land in Victoria. The means are based on data collected over a 20-year period, apparently from the mid-1970s to mid-1990s (Department of Sustainability and Environment 2010a).

Fire cause	Average number of fires each year	Percentage of total fires
Lightning	149	26
Deliberate	145	25
Agricultural	96	16
Campfires	59	10
Cigarettes/matches	41	7
Miscellaneous	26	5
Machinery/exhausts	15	3
Prescribed burn escapes	9	2
Public utilities	7	1
Cause unknown	37	6
Fire cause	Average hectares burnt each year	Percentage of total area burnt
Lightning	53 096	46
Public utilities	16 256	14
Deliberate	15 649	14
Miscellaneous	10 009	9
Agricultural	7799	7
Prescribed burn escapes	5274	5
Machinery/exhausts	2551	2
Campfires	1466	1
Cigarettes/matches	444	>1
Cause unknown	2974	3

order of magnitude from 1 tonne per hectare[8] per year in forests of least productivity to 10 tonnes per hectare per year in forests of greatest productivity (Attiwill *et al.* 1996). This litter-fall includes leaves that have become senescent and died, fine twigs and larger branches, bark and fallen trees. Annual leaf fall across a wide range of eucalypt species is reasonably constant between 2–3 tonnes per hectare per year when total annual litterfall exceeds about 4 tonnes per hectare per year (Figure 3.1). The fall of senescent leaves therefore makes up about 50% of total litterfall when total litterfall is less than 4 tonnes per hectare per year, and decreases to 30% when total litterfall is greater than 8 tonnes per hectare per year.

The plant litter, having fallen to the forest floor, then decomposes. The decomposition constant k can be calculated from the annual rate of litterfall (L) and the quantity of litter on the forest floor (A, in the general range 5 to 25 tonnes per hectare):

$$k = L/A \text{ per year}$$

We can use this simple relationship to calculate the accumulation of litter with time (t)

$$A = L/k\left(1 - e^{-kt}\right)$$

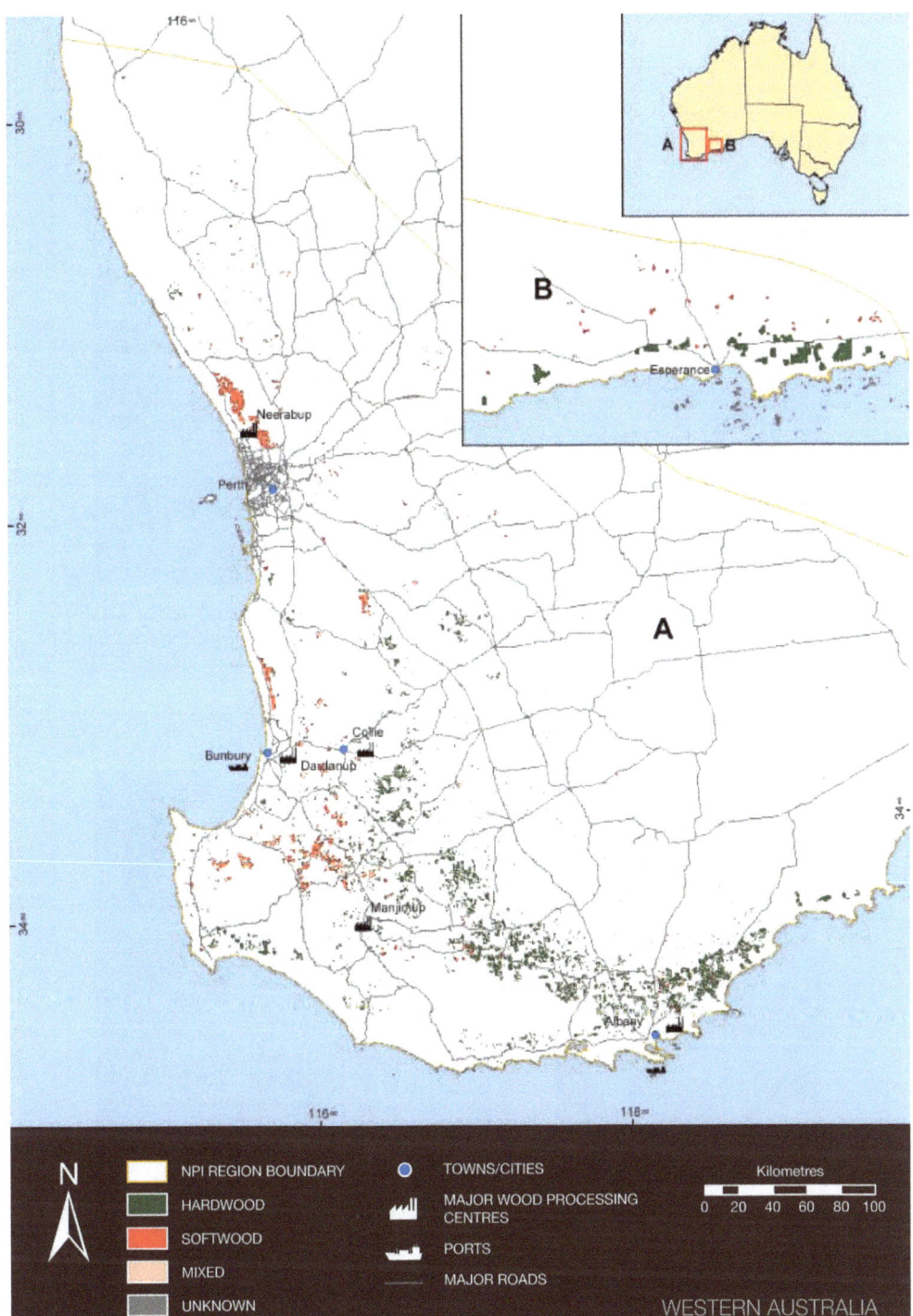

Plate 1. Plantations in south-west Western Australia (from Parsons *et al.* 2006).

Plate 2.   Fire burning with an intensity of 1000 kilowatts per metre; the flame height is 1–2 metres, and the rate of spread is 0.2 kilometres per hour. This intensity is about the limit for suppression by ground crews using hand tools (photo: AG McArthur, Blackmountain, ACT).

Plate 3.   Fire burning with an intensity of 2500 kilowatts per metre; the flame height is 4–6 metres, and the rate of spread is 0.4 kilometres per hour. This intensity is about the limit for suppression by bulldozers and aircraft (photo: NP Cheney, Nowa Nowa, Victoria).

Plate 4.   Fire burning with an intensity of 7500 kilowatts per metre; the flame height is 20–35 metres, and the rate of spread is 1.2 kilometres per hour. This intensity is beyond the limits for direct suppression by any means (photo: J Cutting, McCorkhill Block, WA).

Plate 5.   Fire burning with an intensity of 10 000 kilowatts per metre; the flame height is 40 metres, and the rate of spread is 1.6 kilometres per hour. This intensity is beyond the limits for direct suppression by any means (photo: J Cutting, McCorkhill Block, WA).

Plate 6.   Before (left) and after (right) fuel-reduction fires, Snowy Mountains, New South Wales.

Plate 7.   Whole tree chambers at the Hawkesbury Forest experiment. These allow us to test the effects of rising CO_2 and changes in water availability (rainfall and evaporation) on growth and water use of whole trees (photo courtesy of University of Western Sydney).

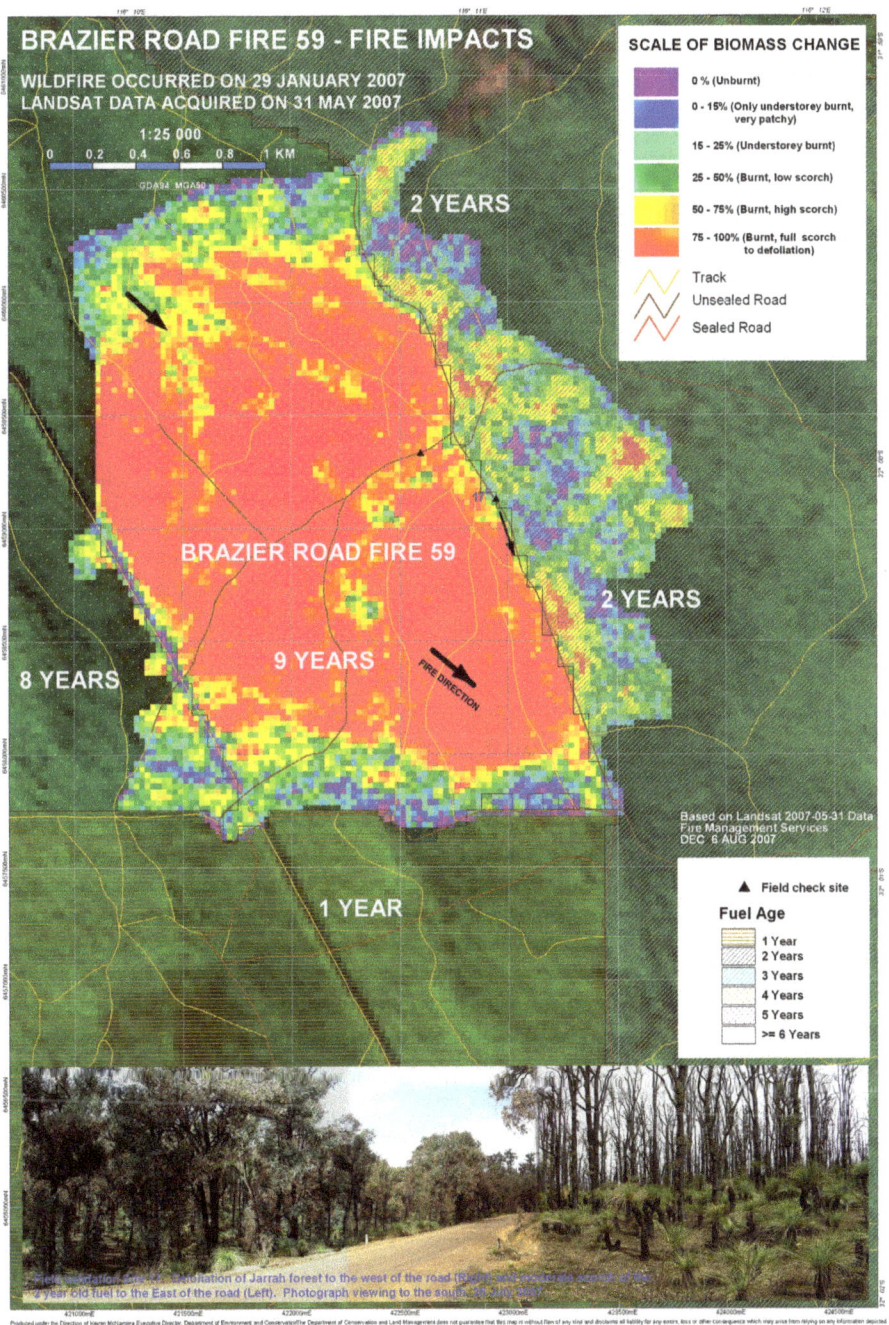

Plate 8. Landsat image showing the extent and intensity of a bushfire, 27 January 2007 in jarrah forest 10 kilometres south-east of Perth during severe conditions. The colours indicate the crown fire that defoliated forest last burned 9 years earlier. The fire rapidly ran into young fuels (1- and 2-year-old) to the south and east where the fire was suppressed as fire intensity moderated significantly. The wide angle photo below shows the difference in the fire impacts in the 9- and 2-year-old fuels to the west and east of the road. (Landsat imagery and photo developed by Dr Li Shu, Department of Environment and Conservation, Western Australia.)

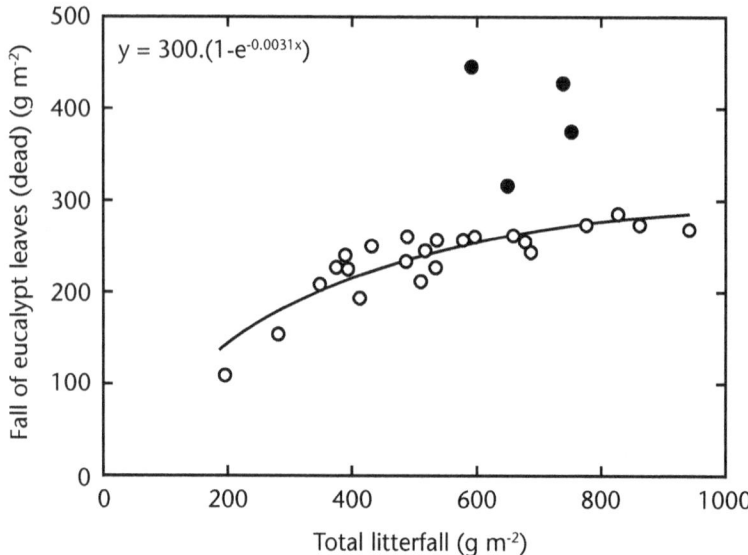

Figure 3.1.   The relationship between the fall of dead eucalypt leaves and total litterfall across a range of eucalypt forests in south-eastern Australia. The four outlying points (closed circles) are *Eucalyptus regnans* (mountain ash) forests; they have not been included in the relationship (from Attiwill *et al.* 1996).

Walker (1981) gives many examples from the literature of decomposition constants for various forests, including calculations of the time ($5/k$) required to reach 99% of the maximum steady state weight of litter on the forest floor. In simple terms, k for many forests is in the range 0.2 to 0.5 per year, so that steady state is reached in the range 10 to 25 years.

Together with ground plants and shrubs, the litter provides abundant fine fuels – up to 30 tonnes per hectare or so – for ignition and spread of fire. The significance of fine fuels is obvious: the finer the fuel, the greater the ratio of surface area to volume, and hence the faster it will burn. In summer, with the fresh crop of newly fallen leaves on the ground, the litter layer of the eucalypt forest can become a tinderbox.

The forward rate of spread of a bushfire is almost directly proportional to the fuel load for given conditions of moisture content and wind speed. These two last variables have major effects on the rate of forward spread. For example, as moisture content decreases from 20% to 7%, the rate of forward spread increases ten-fold, and the rate of forward spread increases exponentially as wind speed increases.

Forest Fire Danger Index

The Forest Fire Danger Index (FFDI) was developed by Alan McArthur (1967). The principal inputs to FFDI are air temperature, relative humidity, rainfall and number of days since rain, a drought index (usually the Keetch-Byram drought index developed for forest fire control in the United States) and wind speed. These data can be entered on a simple, hand-held meter (Forest Fire Danger Meter Mk 5), and the output is an index of fire danger.

Mathematically, FFDI is not a linear function of the variables; it increases exponentially. It is most sensitive to changes in wind speed, secondly to changes in relative humidity, and thirdly to changes in temperature (Dowdy *et al.* 2009).

Box 3.1:   Volatile eucalypts

One of the areas of mystique in relation to eucalypt forests and fire is their 'explosive' combustion, which is usually linked to the release of volatile gases. In more scientific terms, and relative to other plant species, eucalypt leaves do contain relatively large amounts of *volatile organic compounds* (VOCs). The most well-known are the mono- and sesquiterpenoid oils that are stored in subdermal glands (e.g. King *et al.* 2004). Chemically, terpenes are made up principally from multiples of units of the hydrocarbon, isoprene (i.e. nC_5H_8, although there are some oxygen-containing terpenes). Some of the better known among this group of compounds are named after tree species or genera such as pines, camphor laurel and the botanical genus Myrtaceae that contains the eucalypts (e.g. pinene, camphene and myrcene). It is important to recognise that these, and derivatives of these, compounds are continuously released into the atmosphere (not just during fires) by all plants and are usually termed biogenic VOCs. Current knowledge suggests that biogenic VOCs in the atmosphere are dominated by isoprene and rank alongside other greenhouse gases in terms of their effect in interfering with the production of ozone and the creation of aerosols (e.g. Guenther *et al.* 1995). Eucalypts are significant emitters of isoprene (e.g. Street *et al.* 1997; He *et al.* 2000; Loreto and Delfine 2000; Winters *et al.* 2009).

Recent Bushfire CRC studies (Maleknia *et al.* 2008, 2009) of the effects of rising temperatures on emissions of VOCs from eucalypts have highlighted another phenomenon. In addition to terpenes, plants such as eucalypts contain and release considerable quantities of highly volatile compounds including alcohols, ketones and aldehydes. As shown by Winters *et al.* (2009), formaldehyde, acetaldehyde and acetone can comprise a sizeable proportion of the VOC emissions from eucalypts, even at ambient temperatures. In fact, release of these more simple compounds is predictably dictated by temperature. For example, alcohols have low boiling points and are thus released from eucalypts at relatively low temperatures, as shown in Figure 3.2. The rapid release of highly volatile compounds at temperatures much less than flame temperatures (e.g. <100°C) contributes significantly to the flammability of eucalypts, in addition to the slightly less volatile oils. Emissions of alcohols and aldehydes helps explain why the radiant heat from *approaching fires* may cause nearby eucalypts to literally explode.

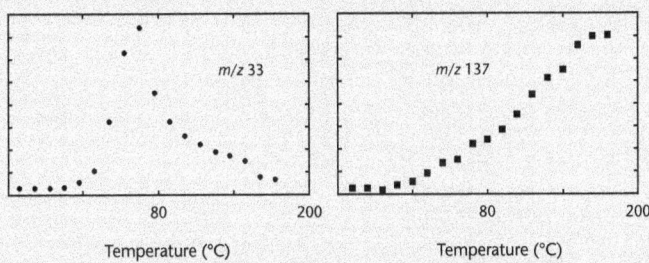

Figure 3.2.   Rate of release of methanol (*m/z* 33) and terpenes (*m/z* 137) from *Eucalyptus grandis* as a function of temperature. The Y-axis shows relative, not absolute, rates (from Maleknia *et al.* 2009).

The FFDI as originally developed by McArthur varied from 1 to 100, the dangers being related to the degree of difficulty of fire suppression:

- **Low <5** – almost no danger of a fire starting, little effort needed by fire-fighters to put the fire out.
- **Moderate 5–12** – moderate possibility of a fire starting, slightly increased effort needed by fire-fighters to put out fire. In other words, on days of low or moderate risk a fire will either go out or be easily put out.
- **High 12–24** – increasing possibility of a fire starting, more effort needed by fire-fighters. Fires can be either put out rapidly or managed in cooler weather in the evening or on the next day. Houses could be threatened.
- **Very High 24–50** – authorities will consider a total fire ban, high chance of fire starting and expanding quickly, fire-fighters need strong efforts. The fire generates sparks and embers, and property can be threatened. Proper safeguards will enable a house to be protected if the householder stays on the scene.
- **Extreme 50–100** – total fire ban, very high chance of fire starting. On these days, a fire becomes uncontrollable very quickly and is impossible to put out, even in low-fuel areas. Fire-fighting resources are stretched to the limit, and the support and preparedness of householders is crucial to the protection of life and property.

Alan McArthur benchmarked the maximum fire danger index of 100 in retrospect from the weather records for Melbourne, Victoria, on the fateful day, Black Friday, 13 January 1939, when bushfires raged through the mountains of Victoria and southern Australia. However, we now know that weather conditions can take the Forest Fire Danger Index well above 100. For example, on Ash Wednesday, 16 February 1983, the FDDI reached 140 in parts of Victoria and South Australia. That was easily surpassed on Black Saturday, 7 February 2009 – the worst natural disaster in Australia's history – when 173 people died in bushfires to the north of Melbourne. David Packham, a meteorologist and fire expert, emailed a number of us a few days before Black Saturday:

> 'The Bureau of Met has issued its estimates for fire weather for Saturday 7 Feb. They are the worst that I have ever seen. The fire intensity calculates at 100.5 megawatt per metre. The max for fire fighting is 2.5 MW/m, crown fires start at about 10 MW/m. I doubt if the State has ever before faced such extreme conditions with fuel levels higher than ever, the prospects for Saturday are horrible.'

In fact, the FFDI on Black Saturday reached 200 and beyond in some parts of Victoria. Following Black Saturday and the disastrous loss of lives, the fire agencies agreed on new categories of extreme fire weather based on the FFDI and the Grass Fire Danger Index (GFDI) (Bureau of Meteorology). The original extreme category (>50) was divided into three levels: Severe, Extreme and Catastrophic (Code Red):

- **Severe:** FFDI/GFDI between 50 and 74
- **Extreme:** FFDI/GFDI between 75 and 99
- **Catastrophic (Code Red):** FFDI/GFDI is 100 or above.

It became quickly apparent that the word 'Catastrophic' was too extreme, too emotive and even perhaps quite misleading. As we finalise this book (October 2010) the Victorian agencies have agreed to drop 'Catastrophic' and use only 'Code Red'.[9]

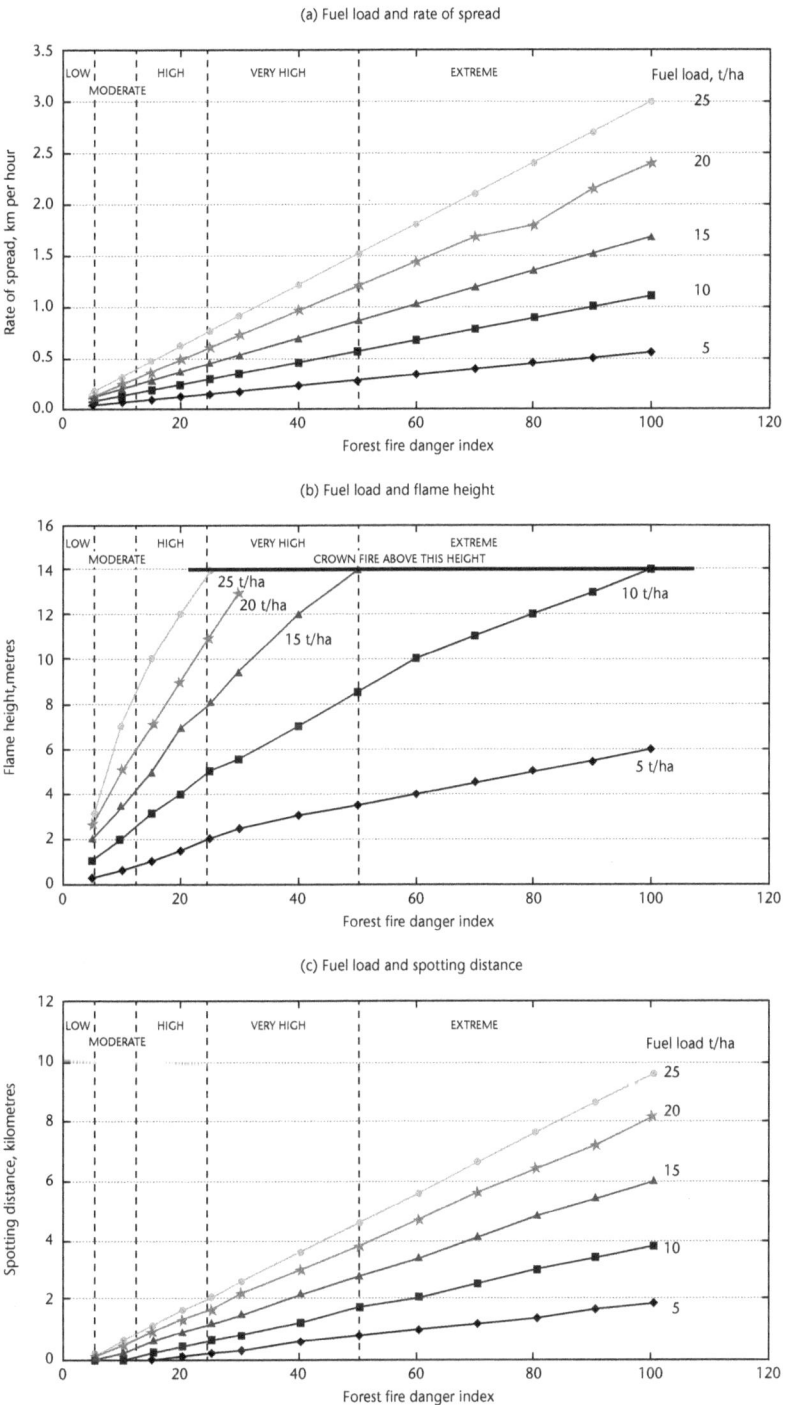

Figure 3.3.   Inter-relationships between McArthur's Forest Fire Danger Index and fuel loads, and their effects on (a) rate of spread of fire, (b) flame height, and (c) spotting distance.[10] The Forest Fire Danger Index is defined as low <5, moderate 5–12, high 12–24, very high 24–50, and extreme 50–100.

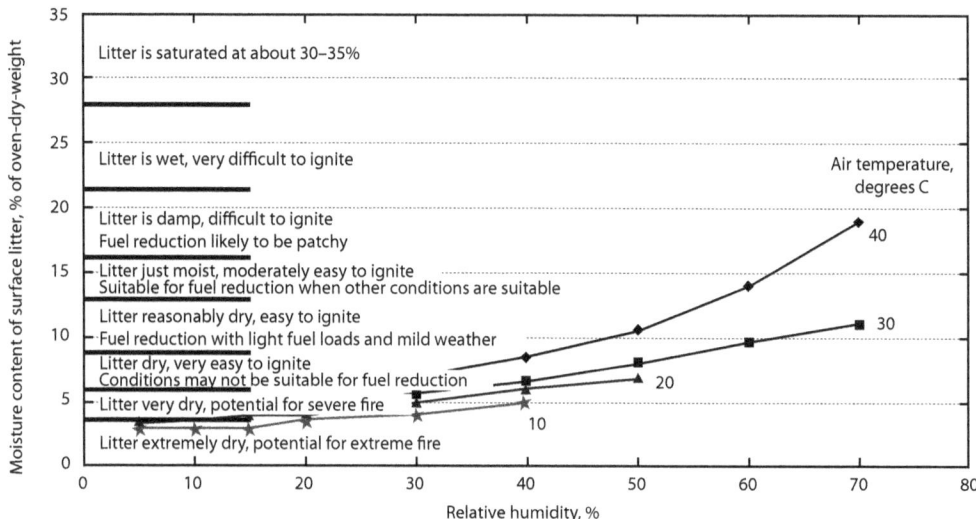

Figure 3.4.   The equilibrium moisture content of surface litter as a function of air temperature and relative humidity (from Luke and McArthur 1978). The moisture content of the litter at a given time depends on rainfall: how much and how recent. The comments on the left-hand side of the graph refer to ease of ignition and fire conditions, depending on the weather (from Tolhurst and Cheney 1999).

The FFDI is based on a quantity of 12.5 tonnes per hectare of fine fuels less than 6 millimetres diameter, and a level to undulating topography. The effects of varying this quantity of fuels on fire properties are shown in Figure 3.3. Both rate of spread (Figure 3.3a) and spotting distance ahead of the fire (Figure 3.3c) increase more or less linearly both with FFDI and with fuel load. However, increases in fuel load rapidly take the flames up into the crowns of the trees (Figure 3.3b) when the FFDI is greater than 20 (in the high range).

The moisture content of the fuel (Figure 3.4) has a very great influence on fire behaviour. Moisture content used to be commonly measured using 'hazard sticks' – wooden sticks that are placed in the forest and weighed periodically to give an *in situ* estimate of moisture content. Moisture content is indexed in the FFDI by the parameters (daily rainfall and maximum temperature) used to calculate the drought index.

The behaviour of a fire depends, of course, not only on weather and fuel but also on the terrain over which it is travelling. Slope and aspect have very great influences on fire behaviour and spotting, and on localised weather events induced by the fire such as the formation of whirlwinds and firestorms. However, this is too specialised for our purposes. The book by Luke and McArthur (1978) remains compulsory reading for those interested in further detail.

The intensity of a fire, or the heat released per unit length of fire edge (Byram 1959), depends on the amount of fuel consumed every second, which in turn depends on the speed of the fire and the amount of fuel available for combustion. Fire intensity (*I*) is calculated as:

$$I = HwR$$

where H is the heat of combustion (joules per gram of fuel), w is the mass of available fuel (grams per square metre) and R is the rate of spread of the fire (metres per second). Intensity, the product of these three variables, therefore has the units joules per second per metre, and since 1 watt = 1 joule per second, intensity (*I*) is given as watts per metre.

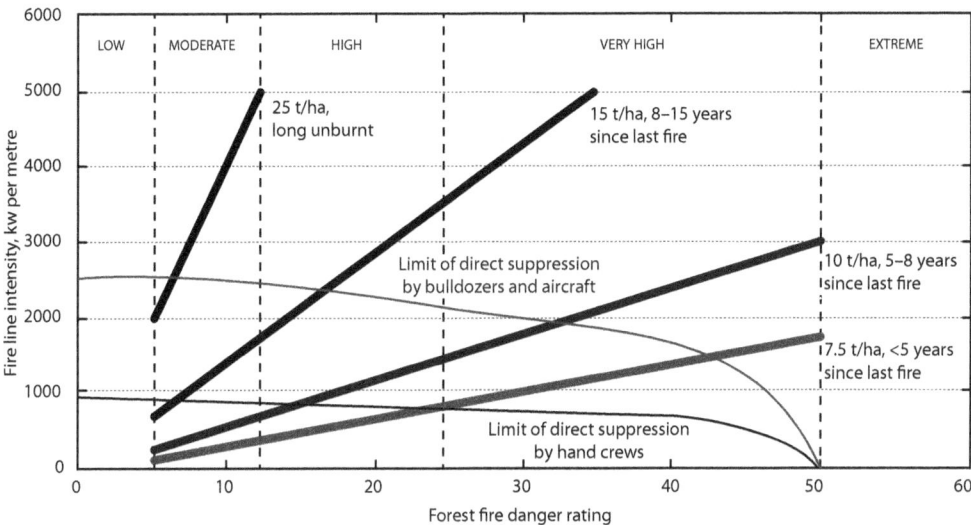

Figure 3.5.   The relationship between fire intensity and Forest Fire Danger Index for a range of fuel loads, and the limits of direct suppression of fire.

The range of intensities for forest fires is enormous – from 20 to 100 000 kilowatts per metre. It is difficult to imagine an intensity of 100 000 kilowatts per metre; however, a large, household radiator emits about 1 kilowatt, so imagine 100 000 large radiators per metre!

At what intensities do we have any hope of controlling a bushfire? (Figure 3.5; from Cheney 2003; and see Colour plates 2–5). A fire can be suppressed by crews working on the ground with hand tools, provided that the fuel load is less than 10 tonnes per hectare and the FFDI is Low to Moderate. Even at the low fuel load of 7.5 tonnes per hectare, ground crews with hand tools are not able to suppress a fire when the FFDI is Very High to Extreme. Bulldozers and aircraft are effective in suppression at the lowest fuel load (Figure 3.3), but they are incapable of dealing with the higher fuel loads. At these higher fuel loads, the intensities are simply too great for direct suppression by any means. In Phil Cheney's words (2004), 'our ability to quell this heat is puny. Under favourable conditions we can put out a fire of 2000 kW/m or 2% of the maximum intensity. This means that under extreme conditions we simply cannot put out a bush fire unless we catch it within minutes of ignition or until it runs out of fuel'.

In short, given the right combination of climate and fuel, bushfires can become so intense over such an extensive area that 'no fire fighting capabilities of any nation' could stop them. Similarly, Noble (1977) quoted Frank Moulds, then Chairman of the Forests Commission of Victoria, giving evidence to a Board of Inquiry into a series of severe fires in 1977: 'At some point in the scale of increasing severity the Commission's fire-protection organisation will be found wanting . . . The community probably could not afford to maintain a fire protection capability sufficient in all circumstances, having regard for the extremes that occur in Victoria'. This argument is taken to the limit by Christensen *et al.* (1989) writing of the huge fires in Yellowstone National Park in 1988: 'Both fire suppression and protection pivot on a paradox. The only way to eliminate wildland fire is to eliminate wildlands'.

4

Ecology, fire and the Australian biota

Ecology: some dominant themes

What is ecology?

Ecology has always been with us, but it was not until the late 1800s that the word was invented. Ecology comes from 'oekologie', derived from the Greek *oikos*, meaning house. Ecology literally means the 'study of the home' – the study of plants and animals (including humans) in the places where they live (their habitat) and their environment.

The environment includes all the physical and biological factors with which an individual interacts and on which it depends for its survival. The organism, in turn, modifies and alters the environment in which it lives. Thus ecology involves the study of relationships between organisms and their environment and the study of interactions between organisms.

The formal definition of ecology follows. Ecology is 'the scientific study of the interrelationships among organisms and between organisms, and between them and all aspects, living and non-living, of their environment' (Allaby 1998).

There is a much broader sense in which the word ecology is used: 'ecology, broadly defined, includes just about everything involving (humans) and (their) environment, and that includes just about everything' (Peters 1991). In this broader sense, 'ecology' is widely used in general conversation and in the media, mostly as an umbrella term that implies knowledge but imparts little scientific information. Michael Allaby (1998) calls this use 'ecologism: 1. The use of ecological terminology or simplistic interpretation of ecological concepts in support of political or moral arguments. 2. Any supposedly ecological expression that is so used'.

And here lies the rub. We are all passionate about our environment and its management. Issues such as old-growth forests, rainforests and fire become overloaded with ecologisms about biodiversity, fragility and ecosystems and so forth that have little or no quantitative basis. This sort of intersection between ecology and politics and moral stances is nowhere more obvious than in the difficult questions that we have to ask about the role and management of fire.

However, the rub is even more basic. There is often disagreement among ecologists and, again, nowhere is that more obvious than in questions of fire and management. There is a wide range of views among ecologists as to the need for prescribed, or planned, burning and, if it is needed, then how frequently and in what season? Although many ecologists, including us, have urged governments to increase the areas of prescribed burns for both ecological and fire control reasons, others are far more conservative about the need for planned fires.

In fact, the science of plant ecology has always had its areas of major controversy. In particular, theories of succession of plant communities were intensively debated over much of the last century. We feel it is worthwhile to outline some of that debate to demonstrate that plant ecology is not always an exact science.

Plant ecology, and succession theory

'Forests often represent a tranquillity associated with the apparent changelessness of large trees' (Shugart and West 1981).

The science that we call plant ecology began with a natural interest in the description of the assemblages of plants within communities. The early plant ecologists were also interested in the ways in which communities changed with time, and these two themes saw the emergence of two mainstreams of plant ecology by the 1920s. In Europe, the Zürich-Montpellier school of phyto-sociology was primarily concerned with spatial change, and so plant communities were classi-fied into smaller and smaller discrete units. In the United States, Frederic Clements (1916) published a monumental work on plant succession and the development of communities. Clements' influence was profound, and it was to dominate ecological thought throughout much of the world, particularly in North America and Britain, through much of the 20th century.

Clements presented a deterministic and holistic view of vegetation change, summarised by the following extracts:

The developmental study of vegetation necessarily rests upon the assumption that the unit or climax formation is an organic entity. As an organism the formation arises, grows, matures, and dies. ... each climax formation is able to reproduce itself, repeating with essential fidelity the stages of its development. ... (The) climax is permanent because of its entire harmony with a stable habitat. It will persist just a long as the climate remains unchanged. ... The life-history of a formation is a complex but definite process, comparable in its chief features with the life-history of an individual plant. All the seres (or successional units) of a climax formation converge to the final community. . . . This fundamental convergence to a climax is developmental, and not individual or local'.

In the simplest terms, Clements believed that there was a succession of changes leading to a stable and harmonious conclusion – a climax that is permanent given that the climate does not change. Clements called this a 'mono-climax' or 'climatic-climax'. Clements' theories have been both supported and challenged vigorously throughout the 20th century. A major step in the United States in the continuing debate over succession was a paper by Drury and Nisbet (1973) in which they emphasised that there are 'few detailed *quantitative* studies . . . . of even short-term changes in vegetation', a remarkable situation, given the length and intensity of concern over succession and climax. Given these limitations, they concluded that changes in structural and functional properties are not consistently associated with changes in species composition, the later stages in succession are not consistently unidirectional, and the effects of species already on the site appear frequently to delay, rather than facilitate, successional replacement.

Ecologists in Australia were lucky not to have been dominated by these overwhelming and strict 'balance of nature' concepts. Nevertheless, much of the teaching still retains major elements of Clements – for example, the notion that wet sclerophyll forests will become rain-forests if left to their own devices. And so we ask the question: why should Clements have held such a dominant position?

The early appeal of Clementsian ecology provided an authoritative statement of pattern and order from which the field ecologist, presented with disorder, could find some comfort

(Attiwill and Wilson 2006). In the wider political and social sense, the ordered progress to a stable, self-perpetuating and harmonious end-point is an overwhelmingly appealing, but unfortunately totally incorrect, view of life. Crawley (1997) wrote that the word 'climax' is so steeped in religious and ethical prejudices (continuous improvement, directed progress towards an ultimate goal, and so on), not to mention Freudian imagery, that its use is probably best avoided. Paul Colinvaux (1993) was much more severe; in his ecology textbook he writes:

'Clements' vision lies at the root of many of the political and social movements that take their names from ecology in the present day. Whenever activists accuse their political or exploiter adversaries of "ecocide" they invoke Clements' teachings. They borrow from him the idea that the ecosystem of the climax is an organism that can be wounded and slaughtered. They are wrong, as Clements was wrong.'

Now we can complete the earlier quote from Shugart and West (1981).

'Forests often represent a tranquillity associated with the apparent changelessness of large trees. Ironically, this tranquillity is largely a product of the human perception of time. . . . Forest ecosystems are dynamic entities that may not be static in either time or space. Many of our concepts of ecosystem management are based on the hope that if a forest is left alone it will gradually return to its natural state. "Succession", "wilderness", "virgin forests" and "climax forests" are all concepts that appeal to the basic notion that forest systems should approach some equilibrium state with time'.

The ecology of disturbance

Over the past 30 years or so, and concomitant with the changing views of succession, it has become widely recognised that plant communities are not in a stable state, living harmoniously with their environment. Rather, plant communities change in space and time, and a major component of this heterogeneity is natural disturbance. An understanding of the role of natural disturbance in moulding community structure and function is therefore fundamental for the management of natural resources.

The recognition of disturbance to a system – as to whether an event is or is not a disturbance – is more or less subjective, depending on the observer's characterisation of the event and on the definition of the system. For example, the heathlands of Mediterranean-type climates are characterised by high plant diversity. Diversity is greatest soon after fire; if fire is excluded, diversity of both plants and animals decreases within 10–20 years (Force 1981; Walker 1982; Fox 1983; Kruger 1983; Specht and Moll 1983) and the heathland becomes dominated by one or a few original or invading species, such as we are seeing in the heathlands at Wilson's Promontory, Victoria (Burrell 1981). To the ecologist, fire in heathlands of Mediterranean-type climates is a natural disturbance – part of the natural order. But to the land manager, protection of life and property may be more immediately demanding than the protection of diversity, and so the manager may treat fire as a catastrophe, not to be allowed.

So we might accept that fire is a natural disturbance, but that a particular fire regime or the cause of a particular fire may not be natural. Fire started by lightning is obviously natural and fire started with malicious intent (i.e. arson) or which has unintentionally escaped from burning-off in neighbouring lands is presumably exogenous (due to forces from outside the community). It is less easy to assign a place on the endogenous (due to forces within)–exogenous continuum of fire, often of stand-replacing area and magnitude, as an integral part of human development over ten thousand years in the Americas (Schule 1990), over tens of thousands of years in Australia (Singh *et al.* 1981; Singh and Geissler 1985) and over 1.5 million years in Africa (Schule 1990). If we were to classify these fires as exogenous, then we deny what seems to us to be obvious – that humans are as much a part of ecosystems as are plants and animals other than humans.

Disturbance is a term that is now deeply embedded in ecology. We have previously outlined the terminology that has developed around the ecology of disturbance, and we have commented on the rather subjective nature of recognising disturbance.

In fact, disturbance events cover a continuum of possibilities. For example, death and decay are normal processes – from the death and fall of a leaf, the breaking and fall of a branch, to the death and fall of a tree. In some forests – many of the tropical forests, for example – seedlings grow and advanced growth is liberated within the gaps created by treefall. Even the destruction of large numbers of trees, blown down by cyclones or hurricanes, provides conditions for the regeneration of the forest. Although we might form the subjective view that cyclones and hurricanes are major disturbances, they too – like death and decay – are part of the natural order of things.

The Australian biota has evolved under a regime of fire. A lightning strike might ignite a single tree. Depending on the weather and the moisture content of the fuels in the forest, the fire started by the lightning strike might simply smoulder and die, or it might burn a few hectares, or it might join up with fires started by neighbouring lightning strikes and burn over large areas. At what stage in this continuum do we view fire as a disturbance, rather than part of the natural order of things?

Insect attack is a major feature of many of the coniferous forests of North America. For example, some 55 million hectares of the north-eastern balsam fir forests were left dead and dying in 1970, and some 23 million hectares were dead and dying following another attack in 1983. While this seems like death and destruction, there is strong evidence that 'spruce budworm outbreaks are periodic natural occurrences that . . . recycle older fir stands to regenerating fir stands' (Ostaff and Maclean 1989). In other words, 'spruce budworm outbreaks are periodic natural occurrences that . . . recycle older fir stands to regenerating fir stands' (Blais 1983; and it is interesting to note here the description of an outbreak as an occurrence, rather than a disturbance).

Aber and Melillo (1991) sum up these important disturbances:

> 'Disturbance by fire, defoliation, or other agents is an intrinsic and necessary part of the function of most terrestrial ecosystems – a mechanism for reversing declining rates of nutrient cycling or relieving stand stagnation. The requirement for fire to reverse soil organic matter accumulation and increase nutrient cycling has been known for some time . . . In the cases of the budworm and the mountain pine beetle, stand break up, the reinitiation of succession, and the reversal of stand stagnation are facilitated by herbivory rather than fire.'

It is interesting to compare the ecological use of the term 'disturbance' with the everyday use. The *Shorter Oxford English Dictionary* defines the verb 'disturb' in terms of agitation, disorder, frustration and even tumult and destruction. 'Disturbance' is then '1. The interruption of tranquillity, peace, rest, or settled condition; agitation. 2. Interruption of mental tranquillity; discomposure. 3. Interference with the due course of any action or process; molestation.'

While we do not want to advocate a change in terminology of disturbance in the well-established ecological literature, it is clear that the ecological meaning of disturbance is rather different from the social or legal meanings. Smaller-scale disturbances such as the fall of branches and the death and fall of trees are not disturbances at all but simply normal and regular events in the life of all plant communities.

And even at the larger scale, the recognition of disturbance is, as we stressed early in this chapter, entirely subjective. Fire has been a major factor in the evolution of Australia's biota over hundreds of millions of years. Because many species depend on the cyclic renewal of resources by fire, how then can fire be viewed as a disturbance? In fact, Vic Jurskis, Forests New South Wales, has argued compellingly that the view of fire as a disturbance is entirely inappropriate for the eucalypt forests of Australia; in total contradiction to that view of fire as a disturbance, Jurskis argues that it is the *exclusion* of fire that should rightly be recognised as a disturbance (Jurskis 2003, 2005a, 2005b).

This contradiction in views emphasises one of the greatest difficulties in the interfaces between ecology, society and politics. Long-standing ecological views of succession as a stable and self-perpetuating state have simply re-enforced the view of fire as a disturbance, with all of its social and political connotations of interference of stability, tumult and destruction. In contrast, fire is an essential part of the lives of our forests. We must accept it, and learn how to manage it so that goals for the management of diversity are achieved.

Fire and the Australian biota

Around the world, fire is recognised as 'the dominant fact of forest history':

> 'The great majority of the forests of the world – excepting only the perpetually wet rain forest, such as that of south-eastern Alaska, the coast of north-western Europe and the wettest belts of the tropics – have been burned over at more or less frequent intervals for many thousands of years' (Spurr and Barnes 1980).

Nowhere is this dominance of ecosystems by fire more important than in Australia. It is often stated that Victoria is the most seriously affected state within Australia, accounting for some three-quarters of deaths and more than half of the economic losses due to bushfire.[11]

Australia was not always like this. Australia, in its northward drift from its Gondwanic connections 130 million years ago, became hotter and drier. The extensive cover of cool-temperate rainforest was gradually displaced by hard-leafed vegetation: by vegetation that, because of its evolution in relative isolation, is highly endemic. With increasing aridity came an increasing incidence and spread of fire from lightning The frequency of fire in Australia increased with the coming of Aborigines some 45 000 to 70 000 years ago, and increased again with the coming of Europeans 220 years ago.

Fire played a dominant role in evolution, an evolution that produced the highly endemic genus *Banksia* and the enormous diversity of species within the genera *Acacia* (wattles) and *Eucalyptus* that now dominate our forests. Many species have characteristics that make them more, rather than less, fire-prone, supporting the hypothesis that 'fire-dependent communities burn more readily than non-fire-dependent communities because natural selection has favoured development of characteristics that make them more flammable' (Mutch 1970).

For example, the leaves of the eucalypts are rich in resins, waxes and volatile oils, the bark of some species is fibrous and stringy and of others, hangs down from the canopy in long ribbons. Many of the forests in higher rainfall areas support a rich understorey, with many species such as *Leptospermum* (ti-trees) and *Melaleuca* (paperbarks) rich in volatile oils, like the eucalypts. Seeds of many species in the genera *Banksia* and *Hakea* are held in hard fruits that open *only* after the heat of a fire.

Many species, both of eucalypts and of the understorey and shrub layers regenerate after fire by epicormic shoots or coppice shoots, or both. Leguminous species (for example, many of the acacias) are hard-seeded; heat apparently cracks the seed-coat, resulting in prolific regeneration after high-intensity fire. Flowering in the Xanthorrhoeaceae (the grass trees) and Orchidaceae and a few other families is fire-induced and gregarious, and the ancient cycad *Macrozamia reidlei* produces cones profusely and gregariously, much like the gregarious flowering of some tropical species after cyclone damage. Smoke from fire initiates and enhances germination of many plant species, and the effect of heat on the soil stimulates the growth of seedlings.[12]

Eucalypts such as *E. diversicolor* (karri) and *E. regnans* (mountain ash) regenerate only after the mature forest has been killed by fire, leaving a bare ash-bed onto which the seeds, liberated from the canopy by the heat of the fire, fall and germinate. For the BBC film 'The private life of plants', Sir David Attenborough (1995) wrote:

'The threat to the spectacular forests of noble mountain ash is not, in fact, fire. It is the absence of fire. If the great trees die from old age before flames have cleared the ground for their seedlings, than they will leave no successors. Paradoxically, such a forest will not survive unless much of it is first destroyed'.

The American historian, Stephen Pyne (1991), wrote lyrically of the eucalypts:

'The Australian bush owes its peculiarity, more than anything else, to Eucalyptus. No other continental forest or woodland is so dominated by a single genus. Other biomes on Earth have scleromorphs, most have grasses, and few are spared wholly from fire, but none has the combination that exists in Australia and has given the bush its indelible character. Eucalyptus is not only the Universal Australian, it is the ideal Australian – versatile, tough, sardonic, contrary, self-mocking, with a deceptive complexity amid the appearance of massive homogeneity; an occupier of disturbed environments; a fire creature'.

Malcolm Gill, CSIRO, proposed a classification of woody species to characterise their response to fire (Gill 1981); this classification is given in simplified form below. The primary level of classification recognises two classes: *seeders* – mature plants are killed by crown fire and regenerate by seed, and *sprouters* – mature plants survive crown fire and regenerate from regenerative buds underground, on the stem, or in the crown. These distinctions are not absolute – many species regenerate both from seed and by resprouting.

1. Plants in the reproductive phase killed by 100% leaf-scorch (non-sprouters, or seeders):
   a) regenerate from seed stored on the plant
   b) regenerate from seed stored in the soil.
2. Plants in the reproductive phase recover after 100% leaf-scorch (sprouters):
   c) regenerate from buds below the soil surface (rhizomes, suckers from roots)
   d) regenerate from aerial buds (epicormic shoots).

A simplified classification for the response of animals to fire is not possible; behavioural patterns and requirements for shelter and food vary greatly among species, and therefore the response of species to fire varies greatly. There are many studies of the responses of birds and mammals to fire, fewer studies of reptiles, and very few of amphibians and invertebrates.

However, of one thing we can be sure: as the cool temperate rainforests that covered Australia began to shrink and were replaced by sclerophyll forests of eucalypts and acacias,

Australia's unique marsupial fauna radiated into the new habitats. 'Accordingly, the fauna may have (*and here we insert our words:* **must have**) evolved in a complex, inter-related system of increasingly dry climate, fire, and a vegetation that enhances and is enhanced by fire' (Catling and Newsome 1981).

Fire, diversity and stability

The greatest concentration of studies of responses of fauna to fire is probably in the jarrah and karri forests of south-west Western Australia, an area of great diversity. In summarising and synthesising these studies, Gordon Friend and Andrew Wayne (2003) concluded:

> '*Fire is an integral part of the ecology of the terrestrial habitats of south-west Western Australia. The mammal species of these ecosystems, like the other constituents of their communities, display a variety of physical and behavioural adaptations that have enabled them to persist in this fire-prone environment. Since the biodiversity and health of these systems are dependent in part on fire, it is not a question of whether contemporary society uses fire as a management tool for conservation, but rather how fire is best used*'.

We have previously used a most important quote:

> '*disturbance by fire . . . is an intrinsic and necessary part of the function of most terrestrial ecosystems – a mechanism for reversing declining rates of nutrient cycling or relieving stand stagnation*' (Aber and Mellilo 1991).

It follows by simple logic that diversity at the landscape level is also dependent on fire at the landscape level. A bushfire at the scale of 1 million hectares, as was the result of the 2002–2003 Alpine fire and the 2007 Great Divide fire in south-eastern Australia, results in less diversity, both in species composition and in vegetation structure within and among communities, than a series of smaller fires over many years.

All of this is nothing new. In 1970, Loucks (1970) proposed the hypothesis that diversity and productivity of forest ecosystems are maintained by random periodic disturbance. There *is* no self-generated (or autogenic) steady state, and the system depends on disturbance of stand-replacing proportions. Loucks ended his paper in the strongest way, concluding that:

> '*The elimination of disturbance by modern humans 'will be the greatest upset of the ecosystem of all time. . . . It is an upset which is moving us unalterably toward decreased diversity and decreased productivity at a time when we can least afford it, and least expect it*'.

We should again differentiate between 'disturbance' as it is used ecologically and in general society. Fire has been a major factor in the evolution of Australia's biota over hundreds of millions of years. Because many species depend on the cyclic renewal of resources by fire, disturbances by stand-replacing fires are simply part of the natural order; it is the elimination of fire that should be more rightly termed a disturbance.

All of this raises many questions. What are the aims – social and biological – of management? What is the level of diversity we should aim for? What levels of disturbance are biologically necessary, and what levels can be socially tolerated? Can we accommodate fire and destruction within the management regime of a national park? These are questions with which land managers around the world have been grappling for a long time. For example, there is now an intensive development of strategies to manage fire as an integral part of the ecosystem in Yellowstone National Park. J.S. Rowe and G.W. Scotter (1973) placed the problem in a general context for the land manager:

'The management of national parks, nature reserves, and wilderness areas poses many questions about the use of fire. The near exclusion of wildfires in such places has had profound effects. If the major goal of such areas is to perpetuate samples of as many landscapes as possible with the recognition that fire is an inseparable part and natural agent in the ecology of many ecosystems, then land managers must "unsell" the false impression that all fires are bad and be prepared to use both prescribed fires and natural lightning fires in landscape management'.

5

Fire and ecological processes

Introduction

Discussions of the use of fuel-reduction fires often go little further than (a) the effectiveness of such fires in reducing the risks of bushfire and (b) the potential effects on biodiversity.

However, in a continent as old and as dry as Australia with a substantial record of Aboriginal burning, there is a *prima facie* case that we should understand the importance of low intensity, moderately frequent fire to factors that make up a substantial proportion of ecological sustainability, namely:

- maintenance of cycles of carbon, water and nutrients
- maintenance of the processes of soil formation.

Cycles of carbon and nutrients fall generally within the field of biogeochemistry. There are several excellent reference works. Beginning with the Bormann and Likens classic (1977), and including Schlesinger (1997), Schulze *et al.* (2001), Melillo *et al.* (2003) and the account by Vitousek (2004), nearly all are written from a northern hemisphere perspective. Perhaps only *Forest Soils and Nutrient Cycles* (Attiwill and Leeper 1987) offers a view from the south.

The field of soil formation is covered by a great many texts, too many to list here. However, one of the important features of Australian soils – the contributions of charcoal (or 'black carbon' or 'pyrogenic carbon') to overall carbon content and other soil properties such as structure and nutrient retention – is seldom covered in detail but it is the subject of much recent research. While public and political attention has drifted toward manufactured 'biochar[13] (e.g. Sohi *et al.* 2009) and converting biomass[14] to long-lived carbon amendments for soils, the more fundamental issues for Australia, given the extent of fires each year (see Figure 2.2), are: 'how much pyrogenic carbon is produced by fuel-reduction fires, what is its chemical nature, and what is its role in soil formation and carbon, water and nutrient cycles?'

In this chapter, we summarise much of what we regard as 'known' about the importance of fuel-reduction fires on water, carbon and nutrients, with particular emphasis on soil aspects. The effects of fires on above-ground attributes of forested ecosystems are mostly obvious and, in large part, a matter of straightforward accounting – how much carbon and nutrients are lost during fires, what is the effect of regrowing vegetation on the amount of water used or on amounts of carbon and nitrogen regained? Important though these may be, they are far more straightforward questions to answer compared with those around soil water, carbon and nutrients.

Fire and carbon

Carbon (C) – an introduction

Sensible discussion of fires and carbon in Australian forests has to start with an appreciation of the global cycle of carbon, with a focus on forests. That appreciation has two key elements.

First, all of the many recent authoritative reports on the carbon cycle (e.g. recent IPCC report by Prentice *et al.* 2001) note that fires contribute very significant amounts of carbon to the atmosphere each year. As shown in Figure 5.1, forest combustion accounts for roughly 4 gigatonnes (Gt)[15] of the carbon entering the atmosphere each year. Consumption of fossil fuels and production of cement account for around 7 Gt each year.

These reports were derived from much prior research, such as the pioneering work by Crutzen and Andreae (1990). Most of the fire-related C emissions are from either tropical forests or from the vast boreal forests and the tundra. As noted by Schiermeier (2005), wild-fires in 22 million hectares of Siberian forests in 2003 wiped out most of the global gains made up to 2012 by the Kyoto Protocol on climate change and kick-started a major research effort to better understand the carbon budget of these forests. Page *et al.* (2002) quantified C emissions from the 1997–08 fires in Indonesia as contributing 13–40% of 'annual emissions from anthropogenic fossil-fuel combustion'. That is, 13–40% of total *global* emissions from fossil fuel consumption for that year were produced by what was effectively a single event in one

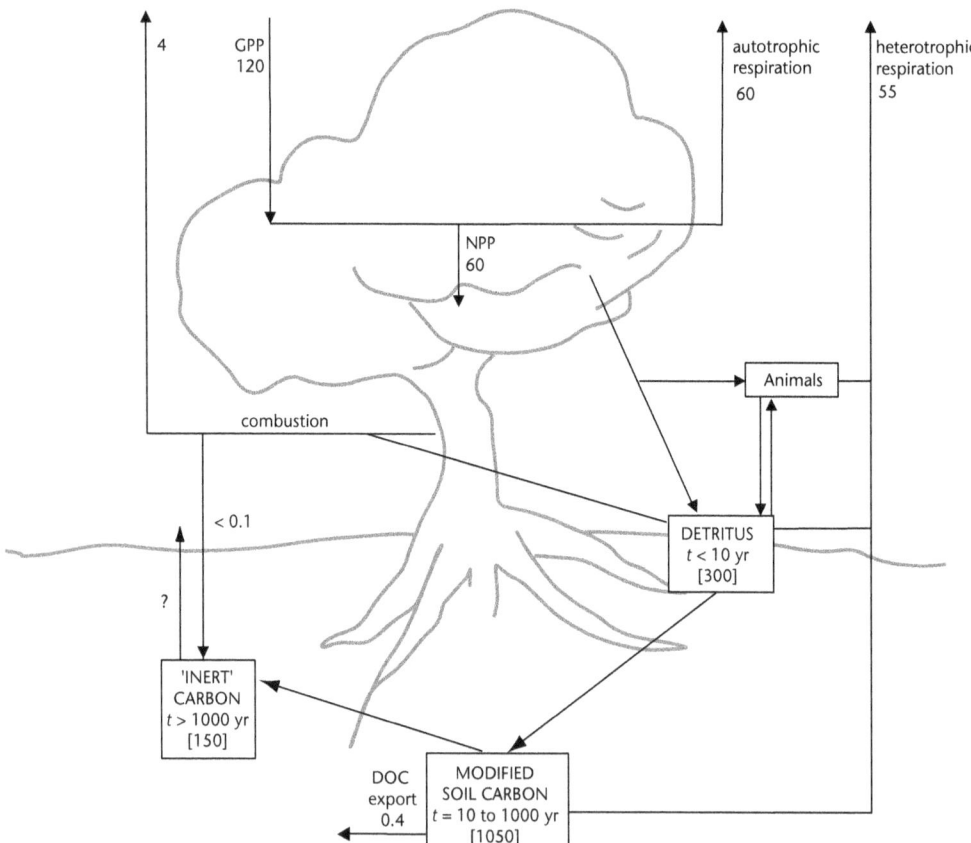

Figure 5.1.   Global carbon cycle (after Prentice *et al.* 2001).

relatively small area of the globe. Schimel and Baker (2002) summarised: '… in areas of high carbon density, catastrophic events affecting small areas can evidently have a huge impact on the global carbon balance'.

Secondly, we now are burning 'old forests', captured in the form of coal, for energy. This is really an intergenerational equity issue – solar energy captured millions of years ago by plants ended up as coal – but we are not doing enough to replace the energy source for future generations. It also nicely illustrates that it is the carbon in soil – not the carbon in trees – that is the long-term (intergenerational) form of carbon storage we need to be most concerned about.

In contrast to the critical importance of soil carbon to discussions about the effects of fires, above-ground carbon is, mostly, a zero sum game – carbon gained is also lost – there is no 'magic pudding'. Sohi *et al.* (2009) state it nicely: 'In fire-affected systems, standing biomass remains in equilibrium, viewed over the long-term'.

The above-ground part of trees provides a temporary storage – if well managed, it might last a couple of hundred years. Forests are *not* continuously increasing sinks for carbon. No matter how vehemently some people might try to portray native forests as eternal sinks, the laws of biology and chemistry still apply. Eucalypts forests sequester C most rapidly in the middle of their lifespan – in southern Australia this might equate with ages 10–100 years. As the forests age further, less and less of Gross Primary Production[16] is retained (more being immediately sent back to the atmosphere) such that Net Primary Production (and the rate of C storage in the biomass) diminishes strongly (see Attiwill 1979). Very old trees are more likely to be net sources of carbon than sinks, notwithstanding contentious claims that old forests continue to show carbon gains, possibly via a shift of production to understorey species (Keith *et al.* 2009). The fact that some of the oldest ash forests (200 years+), growing in some of the most sheltered locations in south-eastern Australia, burnt in 2009, begs the obvious question as to the likelihood of the extraordinary carbon 'stores' proposed by Keith *et al.* (2009).

Practically, the 'old tree' problem is exemplified by their 'hollowing out'. As trees move into old age, fungal and microbial activity ensures heartwood often becomes little more than a vertical 'pipe' of decomposing tissue. Most common among the faster-growing eucalypt species, including mountain and alpine ash, blue gum and mountain grey gum, it helps create the 'roman candles' so often seen during fires, whereby trees burn from the inside out, creating spectacular flame and spark effects.

Effects of fire on soil carbon – processes and storage

Counter-acting forces affect carbon added to soils. Carbon is added directly by plants through litterfall, root sloughing and root exudation, but also indirectly via the remains of decomposer organisms that live on the organic matter added from plants. Carbon is also added indirectly via fires and the creation of char or black C from plants and litter. It is probably more correct to adopt the terminology used by Preston and Schmidt (2006) and refer to 'pyrogenic C' (Pyr C) inasmuch as not all heat-affected or combustion-affected C is actually black.

Respiration, or the conversion of organic matter to carbon dioxide (CO_2), is largely a biological process. Stabilisation, or the conversion of respirable C to more resistant forms, is mostly a geochemical process, although there is evidence that microbes transform some, more oxidised and labile forms of carbon into more reduced and recalcitrant forms. The geochemical process of Pyr C formation during fires is a form of stabilisation.

An example somewhat removed from a discussion of fires may help illustrate the issues. We have known for decades that the ploughing up and planting of pastures with trees results in loss via respiration of a proportion of previously accumulated soil C. Better aeration and the warmer and wetter conditions of recently cultivated soils support the activity of the heterotrophic micro-organisms responsible for the observed increase in respiration. In addition, however, we know that the 'priming effect' – increased respiration of old carbon following addition of 'new' carbon – contributes to the loss. Priming is not yet fully understood but is very significant because more carbon is respired than is added in many cases (Fontaine et al. 2003, 2007).

Together, improved edaphic conditions and priming effects ensure that the carbon balance of a newly established plantation may well be negative for its first few years – that is, more carbon is being sent back to the atmosphere than is accumulated. Eventually, some of the carbon from the trees becomes stabilised via association with clay and soil minerals and the soil recovers what was lost initially. The length of time it takes for this balance to recover is poorly known and strongly site-dependent.

Returning to charcoal and fire-affected detritus, these are stabilised forms of organic matter and highly resistant to respiration and constitute a significant proportion of soil carbon in large areas of Australia. Although no one has yet found terra preta soils (Box 5.1; and see Smith 1999) in Australia, the evidence of widespread Aboriginal burning is comprehensive. Before native vegetation was cleared for agriculture, fires added considerable quantities of charcoal to soils. Agricultural practices such as stubble burning added more. Lehman et al. (2008) reinforced the view that charcoal is a long-term store of C in Australian soils.

Pyrogenic Carbon (Pyr C) and its production during fuel-reduction fires

A straightforward introduction to Pyr C is provided by data from a fuel-reduction fire near Nowra, in coastal New South Wales. After the fire, ash was collected from a range of sites, and sorted according to size classes. Each of the size classes was analysed separately for its carbon and nitrogen content.

The resultant patterns shown in Figure 5.2 are due to:

- the nitrogen (N) content of the material that burnt (a mixture of leaf and understorey shrubs that are richer in N; and woody material that is poorer in N)
- the completeness of combustion (leaves and other fine forms of organic matter are more or less completely combusted, larger woody material is less completely combusted).

As a consequence, carbon concentrations follow a simple pattern (Figure 5.2 Panel A) – there is less C in the fine fractions than in the larger pieces of incompletely combusted woody material. Nitrogen concentrations in the smallest sized classes (Panel A) are low due to more or less complete combustion and the fact that N is volatile at temperatures above 60°C. Nitrogen concentrations reach a maximum in the medium-sized particles and then fall again. This is due to the trade-off between reducing N concentrations with increasing diameter of woody material before fire, and degree of combustion. The low N concentrations of the largest sized particles are as much due to low N content before fire as they are to incomplete combustion. As a result of A and B, the C:N ratio (Panel B) falls quickly as size class increases. This alone indicates a little of why fine ash is such a good 'fertiliser' – it is enriched in mineral nutrients and contains little organic matter.

Beyond such simple analysis, the nature of Pyr C is far from simple. This is currently a large and very active area of research in which Australians have played a significant role. CSIRO scientists Drs Jan Skjemstad, Jeff Baldock and Evelyn Krull have led a range of studies focused on developing methods and techniques for quantifying Pyr C. Their work has been instrumental in allowing others to explore questions such as the role of fires in soil formation and summaries can be found in Krull et al. (2009), Krull et al. (2008) and Skjemstad and Baldock (2007).

Box 5.1: Terra preta soils

'Amazonian Dark Earths (ADE) are a unique type of soils apparently developed between 500 and 9000 years B.P through intense anthropogenic activities such as biomass-burning and high-intensity nutrient additions on pre-Columbian Amerindian settlements that transformed the original soils into Fimic Anthrosols throughout the Brazilian Amazon Basin.' (Solomon et al. 2007).

Also known as terra preta (Portuguese for 'black earth'), ADE have been the subject of intense research in recent years. Mainly due to their remarkably high content of organic carbon, ADE are also rich in nutrients as a result of their formation via low-oxygen combustion of plant material and refuse from human settlements (Marris 2006). As shown in Table 5.1, centuries of smouldering, rather than roaring, fires has created soils that are far less acid than their adjacent counterparts, and enriched in phosphorus, calcium and other elements, as well as carbon.

Table 5.1. Chemical features of Amazonian Dark Earths (ADE) and their adjacent, original counterparts (Org). Data from Liang *et al.* (2006).

Site	Organic C (g kg^{-1})		Total N (g kg^{-1})		pH		Total Ca (mg kg^{-1})		Total P (mg kg^{-1})	
	ADE	Org	ADE	Org	ADE	Org	ADE	Org	ADE	Org
Hatahara	22.0	21.8	1.0	1.6	6.4	4.6	17 545	115	9064	273
Lago Grande	31.5	17.5	1.8	1.3	5.9	4.2	6354	119	5026	251
Acutuba	15.7	15.4	1.0	0.8	5.6	4.7	332	50	777	198
Dona Stella	16.5	10.2	1.1	0.4	5.0	3.9	40	165	139	51

Although clearly a result of intense human activity (as evidenced by pottery and ceramic fragments in the profiles), including the importation of organic matter, and possibly animals, from other parts of the landscape, terra preta soils highlight the possibilities for 'burial' of organic matter (see also Attiwill and Leeper 1987), protecting it from oxidation. They also highlight that nitrogen is much more volatile than either phosphorus or calcium. Such significant changes in soil carbon in such a relatively short period of time, highlight one of the world's most important carbon 'sinks'.

Pyr C produced during fires is highly heterogeneous – probably as heterogeneous as soil C *in toto*. Although its chemical composition is the subject of much recent and highly detailed research, there are a few broad rules of thumb that can be deduced. We have distilled the following rules from recent publications such as those by Baldock and Smernik (2002), Czimczik *et al.* (2002), Certini (2005), Preston and Schmidt (2006), Forbes *et al.* (2006) and Liang *et al.* (2008):

1. Charring is a process of incomplete combustion (between 250°C and 500°C) that, by definition, produces both well combusted and less well combusted material. Classifying

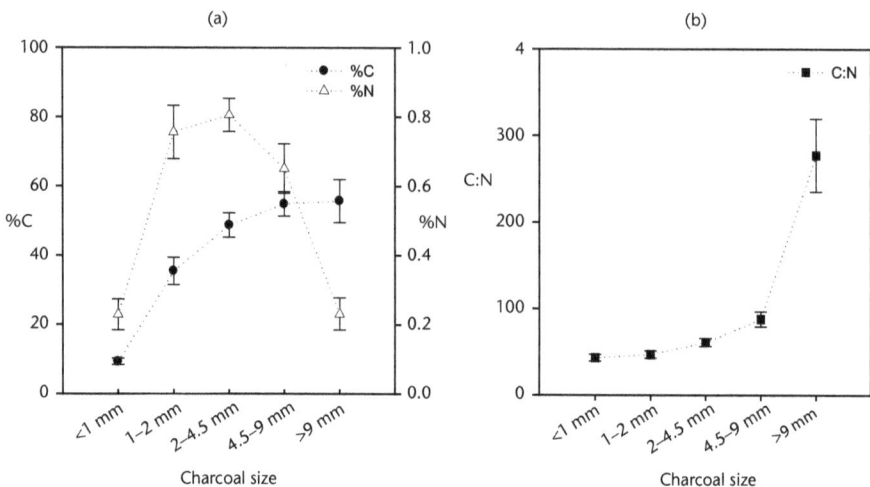

Figure 5.2. Carbon and nitrogen concentrations in ash fractions of different size (Jenkins and Adams, unpublished data).

organic matter as Pyr C or non-Pyr C is an imprecise science owing to the wide range of temperatures in most fires, but especially in fuel-reduction fires.

2. Char contains less of the more labile components (O-alkyl, di-O-alkyl structures) normally found in organic matter (e.g. cellulose and hemi-cellulose) and more recalcitrant compounds such as aryl and O-aryl furan structures. Over 480°C, the character of char is almost completely aromatic.

3. Chemical characterisation of char and of Pyr C in soils are difficult problems. Inter-laboratory comparisons reveal huge variation in quantitative analysis (e.g. Schmidt *et al.* 2001; Hammes *et al.* 2007), especially given the highly sophisticated analytical tools now available, such as nuclear magnetic resonance spectroscopy and ultra-high resolution mass spectroscopy (see Figure 5.3).

4. Consequent to the above uncertainties, estimates of the amounts and decomposability of Pyr C in soils must be treated with caution.

Despite these cautionary remarks, all the available science points to Pyr C formation, and addition thereof to soils, as a potential means of carbon sequestration.

How much Pyr C is created by fuel-reduction fires?

To date, there have been no significant studies of fuel-reduction fires and Pyr C in Australia and international research is minimal. The data that have been collated almost entirely resulted from bushfires or forest regeneration fires. Much of that data was collated by Preston and Schmidt (2006). Rates of *conversion* (usually estimated as the ratio of the mass of charcoal or black or Pyr C after the fire to the mass of litter plus, variously, understorey mass or overstorey mass) are highly variable, dependent on fire intensity, and mostly less than 3%.

Actual masses of carbon deposited as Pyr C on the forest floor ranged from as little as 60 kilograms per hectare of C to about 7000 kilograms per hectare of C. One unusually high estimate of 20 000–170 000 kilograms per hectare of C came from an Australian study (Hopmans *et al.* 2005). In perhaps the single study of a surface fire, Czimczik *et al.* (2005) reported a conversion rate of 0.7% and an increase in soil stocks of Pyr C of around 40% for a Siberian Scots pine forest.

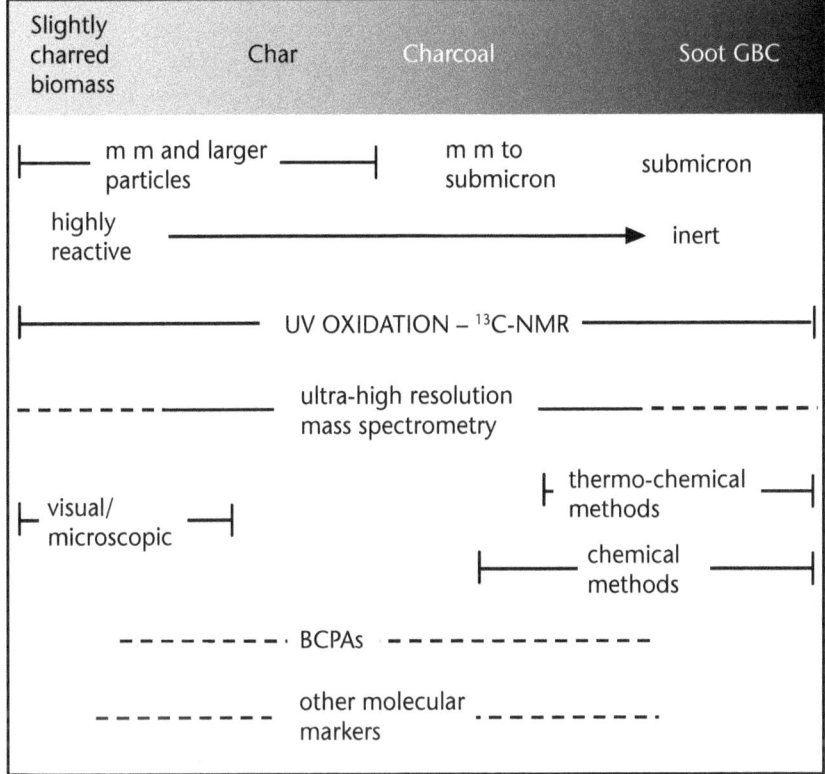

Figure 5.3. Analytical techniques used to quantify forms of pyrogenic C in soils, and their relationship to the size of particles and their reactivity (from Sohi *et al.* 2009).

As an example of just one of the methodological difficulties, charred woody debris is frequently ignored in analysis of production of Pyr C during fires – both bushfires and fuel-reduction fires. Donato *et al.* (2009) recently tackled this issue and developed protocols for quantitative assessment of charred woody detritus.

Notwithstanding the need for further quantitative analysis of Pyr C produced during fuel-reduction fires, the US scientific community has recently emphasised just how important it is through the release of a position paper by the Association for Fire Ecology.[17] That paper concludes with recommendations:

'Based on decades of research and centuries of fire history, we submit that fire, sustainably applied and managed to limit high severity and overall size on landscapes, can be a powerful tool for (1) minimising landscape GHG emissions, (2) increasing ecosystem resilience, and (3) decreasing overall risk of catastrophic vaporisation of carbon stocks that occurs during type conversion.'

The paper goes on to make policy recommendations including these two:

'3) Exempt prescribed burning from annual carbon budgets and caps, due to its critical long-term role in fuels and vegetation management. However, decadal and centenary budgets should be developed for terrestrial ecosystems, and the progress and rate of prescribed burning should be evaluated for ecosystems at these broader time scales.

4) *Evaluate carbon emission and sequestration trade-offs between prescribed burning and wildfires. Carbon sequestration is one of many ecosystem services that can only be measured at decadal and centenary time scales. Only in the long-term does prescribed burning have significant potential to reduce total carbon emissions and increase carbon storage associated with forest burning. Managing more wildfires in remote areas for ecological objectives should also be encouraged.'*

These recommendations are closely aligned with the views of the Wilderness Society, whose scientists Tom DeLuca and Greg Aplet (2008) concluded:

'Charcoal represents a significant component of the soil organic matter pool in temperate grasslands and forests. Charcoal contributes significantly to the total water holding capacity, ion exchange complex, and surface area of the soil environment. Once deposited in soil, charcoal is highly stable having mean residence times 30–100 times longer than that of woody materials and 5–12 times greater than humic materials.'

They concluded further:

'The role of charcoal in the forest ecosystem is just now being explored. The long-term implications of fire exclusion and the elimination of charcoal deposition in forests are not well understood. Timber harvest without prescribed fire may be applied as a forest restoration tool; however, under these conditions charcoal, as a passive C contribution to the soil system, will be eliminated and will lead to a modest, but long term loss of C from the forest ecosystem. Conversely, restoration harvests that incorporate prescribed fire will more effectively emulate natural fire events and deposit charcoal across the activity unit.'

Summary – fire and carbon

There is no doubt that fuel-reduction fires will result in an increase in the amount of pyrogenic C. Although each low-moderate intensity fire will convert only a modest amount of carbon to recalcitrant forms, over long time scales and when summed over the forest estate, this conversion can make a significant contribution to Australia's overall carbon sequestration efforts. Crucial issues for further research include how Pyr C added to soils changes other processes that influence the net balance of soil C:

- oxidation of soil C (see, for example, the divergence of views between Wardle *et al.* (2008) and Lehman and Sohi (2008)
- regrowth of vegetation and its effects on soil C (e.g. Jenkins and Adams 2010), especially through the 'priming effect'
- emissions of other greenhouse gases such as methane and nitrous oxide.

Fire and nutrients

Nutrients – an introduction

Nutrient limitations, especially phosphorus limitation, are important themes in studies of nutrient cycling (Attiwill and Adams 1993). In this respect, the age of any given soil is highly relevant because we know that, with the passage of time, key major plant nutrients such as calcium, magnesium, potassium and phosphorus, are depleted by leaching (e.g. Walker and Syers 1976), while nitrogen generally accumulates. For most of Australia's southern forests that have very old soils, we might expect shortages of phosphorus (P), potassium (K), calcium (Ca) and magnesium (Mg), at least relative to nitrogen (N), with the passage of time. We need to be careful with terminology because there is no evidence that plants in native forests are nutrient deficient. They might be limited in their growth as a result of poor supplies of P or K, or any of a variety of other elements, but by regulating growth according to nutrient availability, they avoid suffering from true nutrient deficiencies.

Fire has been a companion to the evolution of the flora of southern forests. On a global scale, 'wildfire began soon after the appearance of terrestrial plants in the Silurian (420 million years ago)' (Bowman *et al.* 2009). As the Australian continent has changed from being colder and wetter to its current warmer and drier condition (Hill 2004), fire has become more prevalent. Although southern eucalypt forests occupy the 'wetter' parts of the continent as a whole, on a global scale they are far from 'wet' and, for the most part, have to tolerate Mediterranean-like conditions, especially many hot dry months when rainfall is exceeded by open-pan evaporation.

Like Australia, large areas of the African continent are dominated by water-limited ecosystems. These two continents share an abundance of fire-prone and geologically and geomorphically old and highly weathered landscapes – landscapes that are more widespread in the southern hemisphere continents than they are in Europe or on the North American continent. The 'fires and nutrients' view from the 'south' has thus always been different (e.g. Adams *et al.* 2004; Adams 2007) to the nitrogen-centric view from the 'north', where water and phosphorus are usually at least adequate for growth of many species. 'Northerners' have also focused on nitrogen as far as forest management is concerned, because of atmospheric pollution, which ensured consistently large inputs of nitrogen, often in excess of plant demand, resulting in nitrate pollution of natural waters (e.g. Aber *et al.* 2003). That viewpoint is changing such that there is better global understanding of the limiting nature of finite supplies of phosphorus and that this can have significant repercussions for species survival and distribution in parts of the northern hemisphere (e.g. Wassen *et al.* 2005), as well as in the south.

It is obvious that fire goes hand-in-hand with either long-term or seasonal shortages of water in many of the world's ecosystems (or at least it did prior to European intervention), as it does in Australia. The following discussion of fire and nutrients, focused on fuel-reduction fires, thus draws on global, as well as Australian, literature.

Perhaps the most important principle in discussions of fires and nutrients is that it is the amounts of nutrients 'in circulation' (or cycling) that matter to plants, much more than amounts locked up in insoluble forms in soil, in the large woody debris on the forest floor or in the heartwood of trees. These features of nutrient cycling in all forests worldwide are well covered in the texts mentioned in the Introduction to this chapter.

Nutrition of Australian forests in a global context

One aspect that has drawn repeated attention from ecologists is that nutrients in the soil must be available in proportions to each other that resemble their proportions within plants. Ratios of one element to another in plant tissues lie within narrow ranges. Known as Redfield ratios after the pioneer of stoichiometric analysis in marine systems (Redfield 1958), they have been widely studied.

An outcome of these studies (e.g. Aerts and Chapin 2000; Koerselman and Meuleman 1996) arises from observations that within functional groups (e.g. heathland plants or trees), N:P ratios are clustered around a mean of 15. Koerselman and Meuleman (1996) concluded

Table 5.2. Mean N:P ratios in foliage for selected dominant genera in three communities at Anglesea, coastal southern Victoria. Also shown for each community are the mean proportional withdrawals of N and P during leaf senescence (from Taranto 2003).

Genus	Bald hills heathland		Heathy woodland		Closed shrubland	
	Mean (SE)	n	Mean (SE)	n	Mean (SE)	n
Eucalyptus	27 (0.7)	36	29 (0.9)	45		
Gahnia	37 (1.3)	36	37 (1.8)	35	36 (1.4)	33
Melaleuca	–	–	–	–	40 (1.0)	35
N resorption (%)	58		57		53	
P resorption (%)	92		87		80	

plant growth was limited by P when N:P was greater than16 and by N when N:P was less than 14. As an example, Keay and Bettenay (1969) collated data from a study of over 60 species from a range of arid heathlands and eucalypt woodlands in Western Australia, and showed that, of the 69 species investigated, 56 species had N:P ratios often very much greater than 16. Wider analysis of Australian forests suggests the same dominance by plants with N:P very much greater than 16.

This pattern holds, more or less, for all native plants in southern eucalypt forests. A useful case study combines heathlands with heathy woodlands and shrublands in a near-coast environment in south-east Australia (Taranto 2003). Soils are uniformly poor in this area. Tops of hills are dominated by heathy woodlands, heathlands dominate mid-slopes and valley bottoms are covered by closed shrublands. A range of parameters were compared among the three communities using, wherever possible, the same species as the basis for comparison (Table 5.2). Identical species of *Eucalyptus* (tree) and *Gahnia* (sedge/herb) were common in the first two communities and *Eucalyptus* was replaced by *Melaleuca* in the closed shrubland.

Three points are immediately clear. The N:P ratios of all species were well in excess of 16. Secondly, the lowest N:P ratios were observed for emergent trees – an expression of within-community variation in N:P caused by large individuals within a small-stature community. The big trees have better access to resources than the small shrubs and herbaceous species (Kerkhoff and Enquist 2006). Finally, the native species show strong ability to withdraw N, and especially P, from foliage before it is shed as litter – this is a crucial feature and widely interpreted as one of the key mechanisms of adaptation to low availability of nutrients. Given the generally low nutrient status of soils supporting forests and woodlands in Australia, N:P ratios much greater than 16 are consistent with the long-held hypothesis that low P status is an abiding feature.

Only grasses and forbs in southern eucalypt forests ever approach an N:P of 16 – the availability of P ensures that the window of opportunity for these species is small – restricted to the first few years after fire.

When the global data are considered, strong patterns emerge (Han *et al.* 2005; Kerkhoff and Enquist 2006; McGroddy *et al.* 2004; Reich and Oleksyn 2004). These syntheses show that:

1. the N:P ratio of foliage increases with mean annual temperature (i.e. toward the equator, Reich and Oleksyn 2004: see Figure 5.4. Lines A–D)
2. the N:P ratio of total plant mass is invariant to changes in plant size (across three orders of magnitude in plant size, Kerkhoff and Enquist 2006)

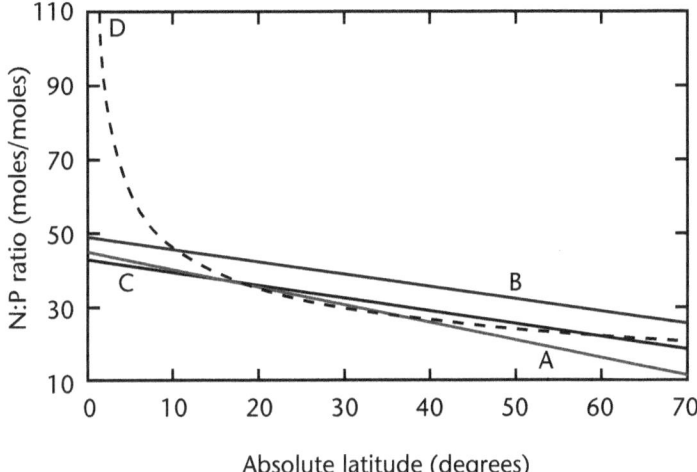

Figure 5.4.   N:P ratios of foliage and leaf litterfall as assessed by different research groups. For foliage, Line A = Reich and Oleksyn (2004); Line B = Kerkhoff and Enquist (2006); Line C = McGroddy *et al.* (2004). For leaf litterfall, Line D = McGroddy *et al.* (2004). (After Hedin *et al.* 2004.)

3. because productivity is allometrically related with biomass (or phytomass), productivity per unit N or P actually declines with increasing biomass (Kerkoff and Enquist 2006).

These patterns owe much to the distribution of soils, especially soils of different ages, on a global scale. The older soils common to the tropics are highly weathered (as a result of greater rainfall and temperature) and are low in P. Species adapted to these soils withdraw much more P than N as the foliage senesces and becomes litterfall (e.g. Line D). Soils elsewhere are less weathered and are generally N-limited. The Australian continent has drifted northwards and the soils are overwhelmingly old and have lost much of their original P. When these factors are considered, Australian soils, and more particularly Australian forest vegetation, fits nutrient patterns established for all plants and forests. As an example, the data shown in Table 5.2 suggest that leaf litterfall in these plant communities will have N:P ratios of between 40 and 60 – well above the average ratios for plants at similar latitudes elsewhere, but in keeping with patterns for P poor soils that are more common in the tropics.

Fires and nutrient losses and gains

In relation to fire, especially fuel-reduction fires, people usually think first of the possibility that fires cause losses of nutrients and, perhaps second, of the enhanced growth after fire of many native plants ('the ashbed effect'; Humphreys and Craig 1981; Chambers and Attiwill 1994). The processes that add nutrients to ecosystems in the longer term, such as biological nitrogen fixation (Adams and Attiwill 1984a, 1984b; Attiwill and May 2001; May and Attiwill 2003; Hamilton *et al.* 1991) and inputs of other elements from the atmosphere in rainfall and dust (e.g. Adams and Attiwill 1986), are more seldom considered.

The literature with respect to nutrients and fuel-reduction fires can be confusing. It can be, and has been, interpreted by some as containing warnings about the risks of fuel-reduction fires due to nutrient losses. Yet these risks are so remote under the current, and any likely future, scenario of fuel reduction at the landscape scale as to be of minor consequence. Much of the confusion can be removed by clear reference to the principles of chemistry and physics, and to the evidence in the forests themselves. The principles of physics and chemistry govern the actual losses of nutrients contained in the litter layer and understorey, and how much is returned to the soil, albeit sometimes in different forms to those in the soil before fire. Here at least, things can be simplified (see Attiwill and Leeper 1987). Nitrogen is the most volatile of the major nutrients and is quickly lost from organic matter once temperatures rise above 100°C or so. The other major plant nutrients are only volatilised at much higher temperatures (e.g. over 500°C for P, Boerner 1982).

During the 1980s there was considerable debate about the loss of nutrients due to fire; much of this debate focused on phosphorus (P), and it was stimulated by the forestry practice of clear-felling followed by burning of the logging debris to achieve regeneration. John Raison and his colleagues from CSIRO in Canberra argued that fire will lead to nutrient depletion and hence to a loss of productivity (Raison 1980; Raison *et al.* 1985a, 1985b). Pragmatically, however, if fires pose significant risks of losses of P then, for example, the forests around Darwin in the north should be depauperate given the frequency of fires. Likewise, the recorded and quite astonishing productivity of mountain ash forests of the south after fires of the greatest possible intensity, could not be countenanced if P was lost at anything like the rates suggested by studies of Raison *et al.* of fires of much lesser intensity.

Part of the difficulty is establishing the reference against which to measure losses during fuel-reduction fires. We have noted elsewhere (Attiwill and Adams 2008) that choosing long unburnt forests as the reference creates immediate issues for analysis of the effects of fires on water yield. It also creates issues for assessment of nutrient losses. Obviously, the longer the period without fire, the greater the amounts of nutrients contained in litter layers, especially nitrogen but also phosphorus.

On the other hand, there is no doubt that cool–moderate temperatures of fuel-reduction fires (see Chapter 3) change the balance of availabilities of one nutrient to another in the soil – nitrogen is truly lost, as is a proportion of the K and Mg, but by the time we consider Ca, and especially P, little is ever lost. Instead, ash is rich in P, Ca and Mg. Before fire, nitrogen is freely available as a result of N-cycling processes but little P is available because it is increasingly locked up in soils and litter (e.g. O'Connell and Mendham 2004). Apart from a short-lived increase in the pools of available N (see also the meta-analysis of the effects of fires on N by Wan *et al.* 2001), fuel-reduction fires mostly reduce the amounts of N in circulation and *increase* the amounts of P, Ca and Mg. As demonstrated by Hungate *et al.* (2003b), although many biological processes in ecosystems have little effect on the stoichiometry (e.g. C:N, N:P or C:P) of the major nutrient elements, fires change those ratios substantially owing to the differing volatility of elements.

Leguminous species play a crucial role in promoting regrowth after fire by replacement of lost N and it is here that we believe resolve the arguments around nutrient losses and replace-ment. For how long should we wait to ensure N lost during fire is replaced by biological nitrogen fixation? Fortunately, the enhanced P status of soils post-fire encourages biological nitrogen fixation such that its rate is usually greatest within a year or two and then slowly declines (e.g. Pfautsch *et al.* 2009). Periods of 5–10 years between fuel-reduction fires seem likely to be sufficient to replace the N lost.

Apart from the magnitude of the loss and the rate of replacement of nutrients following clear-felling, with or without burning, the question remains as to whether or not there is an ensuing loss of productivity of the forest. At the height of the debate over nutrient losses due to fire and logging, Stewart *et al.* (1985) noted that 'no study has demonstrated a decline in the productivity of natural eucalypt forests managed primarily for timber production' and they added pragmatically that 'E. regnans forests are highly productive despite having endured significant nutrient loss during and following severe wildfires at frequencies of 200–300 years'.

A way of testing the hypothesis – that the loss of nutrient following timber harvesting causes a loss of productivity – is with a 'reverse-depletion' experiment (Stone 1979). That is, if a loss of 100 kilograms per hectare of N is to give a decrease in growth, it can be argued that an addition of more than 100 kilograms per hectare of N will give an increase in growth.

Fortuitously, the fire of Ash Wednesday, 16 February 1983, in the highly productive mountain ash forest at Britannia Creek, Victoria (see Preface, Box 1) gave us the chance to test whether or not there is a loss of productivity. Logging over the 1982–83 summer at Britannia Creek had almost finished when the Ash Wednesday fire burnt through some 40 000 hectares of forest. It burnt a just-completed logging coupe of some 40 hectares with great intensity, so that nutrient losses by the combined effects of logging and burning should have been very large. We established a fully-replicated reverse-depletion experiment, and we have been measuring it ever since. There has been no response in the rate of basal area growth to additions of P up to 500 kilograms per hectare, of N up to 1000 kilograms per hectare, of N and P added together, and of N and P added together with all other essential nutrients (Attiwill 1992, 1994, Attiwill and May 2001). This result does not support the hypothesis that the loss of nutrients caused by harvesting results in a loss of productivity of the forest. The result is strengthened by the fact that the forests at Britannia Creek have been intensively harvested both for sawlogs and for wood distillation since the turn of the century; the coupe on which the reverse-depletion experiment is based, and which is sustaining high rates of productivity, had been logged by clear-felling some 60 years previously.

In summary, we concur with the excellent review of fire effects on forest soils by Certini (2005): 'despite common perceptions, if plants succeed in promptly recolonising the burnt area, the pre-fire level of most properties can be recovered and even enhanced'.

Fires, phosphorus and species composition of southern forests

Following from the above, chronosequence studies (Wardle et al. 2004) in Australia and New Zealand suggest that P limitation, developed through accumulation of P in recalcitrant woody biomass and detritus (e.g. Richardson et al. 2004, 2005), is strongly indicated in the collapse, or potential collapse, of forests. Unless there are disturbances that refresh and renew phosphorus supplies in the soil, ecosystems stagnate, productivity declines and dramatic changes in species composition follow. Many people would argue that this is happening in large areas of land in southern Australia, especially land that is now managed as national parks.

Wardle et al. (2004) proposed a most important hypothesis in an Australian context: a hypothesis that rings true for many Australian ecosystems – vanishingly small pools of P in soils are locked away over time in occluded forms in soil or in plant matter.

We believe there is another strand of evidence that links fires, phosphorus and major changes in species composition. William Bond and others (2004) drew attention to the role of fire in plant community distribution at the global scale. Bond's work, based largely on African examples, showed what the world's vegetation patterns might look like with or without fire. The overwhelming evidence is that in the absence of fire, woody shrubs and trees would replace grasses and forbs, and thus savanna grasslands will become woodlands. This evidence accords with the evidence of P limitation and N:P stoichiometry as advanced above, and with Wardle's hypothesis. Grasses and forbs require N and P to be available in different ratios (close to 15:1) to that required (or which can be tolerated) by woody shrubs and trees. In water-limited ecosystems on old soils, only after fire does P availability increase to be sufficient to meet the needs of herbaceous species. Wardle's hypothesis – that reductions in phosphorus availability and cycling are inevitable without disturbance and are accompanied by losses of species diversity and productivity – implies that this situation will become worse as the soils age.

The effects of fire regimes on the nutrient status of forests have long been argued from all sorts of perspectives. In addition to the reluctance by some states to use fuel-reduction fires on the basis of possible nutrient losses, McIntosh *et al.* (2005) concluded that fires may play a significant role in the physical development of texture-contrast soils (common in lower rainfall areas) because 'fire will encourage clay eluviation'. These authors endorsed a previously advocated 'feedback mechanism' (as proposed by Bowman *et al.* 1986, and Jackson 2000) that 'a fire-prone ecosystem proceeds irreversibly down a pathway of incremental nutrient loss and increasing susceptibility to further fires, because the decreasing nutrient status of the ecosystem encourages fire-tolerant forest communities'. As we have noted above, if fires cause such losses of nutrients then much of the Australian forest estate would not be forest at all, but instead would be desert. In addition, we know of no evidence that links clay eluviation to fires, nor do we know of definitive evidence that supports fire-tolerant communities replacing fire-intolerant communities, other than changes in climate (and then mostly on geological scales). Likewise, 'nutrient status' in respect of plants, means the availability of nutrients and this is determined, as we state at the outset of this section, by the amounts in circulation, and cannot be simply assessed on the basis of a soil test. In the mostly water-limited forests of southern Australia, the amounts of nutrients in circulation, and thus available, are due to plant growth and organic matter turnover and are regulated by water – there is no significantly large 'pool' of nutrients sitting idle waiting to be drawn upon by plants. The 'downward spiral' argument – fires creating nutrient losses that in turn promote fire-tolerant communities – seems to us to be rather overstated given the critical importance of fire in *mobilising key nutrients such as P that otherwise become locked up in senescent organic matter.*

As an example of the roles of water and water availability in P cycling, we can consider a recent study of Mediterranean oak. Drought produced negative impacts for P cycling (e.g. Sardans and Peñuelas 2004) that are likely to exacerbate losses of P from the active pools. Chief among the effects were a substantial increase in the increase of P in litter layers (as a result of leaf shedding), and a reduction in canopy P (from the same cause). Enhanced additions of organic P to litter layers can take many years to filter back to the active pools of P owing to extensive P immobilisation in organic matter in soils and litter already poor in P.

In contrast to the 'downward nutrient spiral' ideas of McIntosh, Bowman and others, Vic Jurskis and colleagues (e.g. Jurskis 2000, 2005a, 2005b; Jurskis and Turner 2002) have instead put forward ideas that are more in accordance with worldwide knowledge of nutrient cycling and with known chemical and physical principles. They have used their work to call for restoration of more frequent low intensity fire as a solution to instances of tree and forest decline throughout southern Australia.

Jurskis and colleagues have attributed forest decline to a build-up of nitrogen in the absence of fire. This accords with known patterns of accumulation of N in all ecosystems worldwide. In an Australian context, and as outlined above, the Jurskis hypothesis might well be rephrased as the build-up of N causes an imbalance in N and P availability – too much N, to go with too little P. The Jurskis model and the McIntosh model are drawn with that of Wardle *et al.* (2004) in Figure 5.5. The Jurskis model is at least partly supported by other studies across southern Australia, including those in the Tuart forests of Western Australia (see Close *et al.* 2009).

The weight of global evidence supports the proposition that in the absence of fire, N:P ratios will increase. A widening N:P ratio can cause large, even wholesale, changes in diversity (e.g. change from grassland to heathland or forest) and productivity. If we wish to maintain grassy understoreys in some forests, and ensure the conservation of the great many grass and

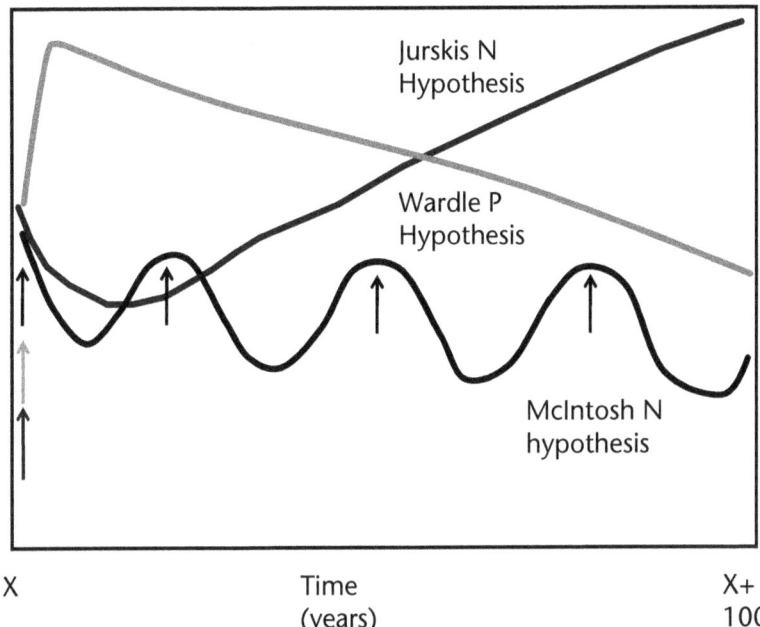

Amounts of nutrients (N or P) in circulation and thus available to vegetation

X

Time (years)

X+ 100

Figure 5.5.   Three possible interpretations of significant changes in nutrient availability due to fires. The 'Jurskis' and 'Wardle' hypotheses are focused on likely effects of long periods without fire (e.g. 100+ years) such that either N accumulates or P is immobilised resulting in significant differences from initial state. The 'McIntosh' hypothesis suggests continual losses of nutrients with each successive fire. See text for further details.

herb species that are endemic and restricted to our native forests, then more rather than less frequent fires will be required. The alternative is one of dense woody shrub understoreys and, according to Wardle *et al.* (2004), eventual ecosystem stagnation and collapse.

Summary – nutrients and fuel-reduction fires

Vitousek and Howarth (1991) posed the rhetorical question as to how nitrogen limitation could ever develop in a world where biological nitrogen fixation is abundant. The question needs posing in the north because nitrogen still appears the most likely candidate for a nutrient limitation, yet N-fixation is so pervasive it seems hard to imagine there could ever be insuffi cient N to meet plant demands. The same applies in southern forests in Australia. N-fixing species and N fixation are ubiquitous.

For southern forests, fire provides the answer to Vitousek's rhetorical question. Without fire, most sclerophyllous Australian forests might never be truly nitrogen limited – but with fire it seems axiomatic they could be, at least for short periods. Even here though, P plays a critical role because, without the mobilising effects of fires, it seems unlikely that there would ever be sufficient P to meet the demands of N-fixation that are notoriously P-demanding. Eventually, however, P will take over again from N as the most limiting of the major nutrients.

We are arguing that shortages of plant-available P must be included in our view of fuel-reduction fires in southern forests. There is strong evidence in extremes of observed global ranges in key measures of nutrient efficiency and productivity among the plant species endemic to our forests that P is the element of most concern. Nevertheless, plant evolution has ensured that if the supplies of P are too low for one species, there is another species better able to make use of what there is. Eventually, though, there are very few species that can cope with vanishingly small supplies of P and ecosystems stagnate and become rather species-poor. Fire is essential to refresh P-cycling, species diversity and productivity.

Fire and water

Introduction – fires and water in mountain ash forests

The hydrology of forested catchments in Australia is best known for mountain ash (*Eucalyptus regnans*) and it seems appropriate to begin there.

We have known for decades that high-intensity fires have dramatic effects on water yield from forest. For example, Figure 5.6 is a version of the 'Kuczera curve' for mountain ash (*Eucalyptus regnans*) forests near Melbourne. It shows that forests regrowing after bushfire use more water in transpiration, and thus yield less in streamflow. In some cases, the yield increases in the first year or two after fire, depending on how quickly the regrowing forest can get established and the timing of the fire. But, overall, the pattern is well established.

Several caveats need be applied. First, 200-year-old *E. regnans* is a rarity because of fire. As an example, the recent 2009 fires in Victoria wiped out the old-growth mountain ash forest in the Yan Yean catchment at Wallaby Creek that contained trees of 300+ years of age. The fires also burnt nearly all the mountain ash forests in the O'Shannassy and most of the Maroondah Catchments. By far the majority of the rest of the *E. regnans* forest estate in the water catchments is less than 100 years old and is regrowth after major fires, such as those in 1912, 1926, 1939, 1983 and now 2009.

If we were to use an 80-year-old forest as the most sensible reference (and much of the catchment forests are actually 70 years old or less) we might well focus our management on the security of water yield, and accept that catchments need to be a mosaic of different ages – rather than the current 'boom and bust' style of management (and simply hoping that catchment forests will reach 200 years of age) that transfers the problem of water yield to future generations.

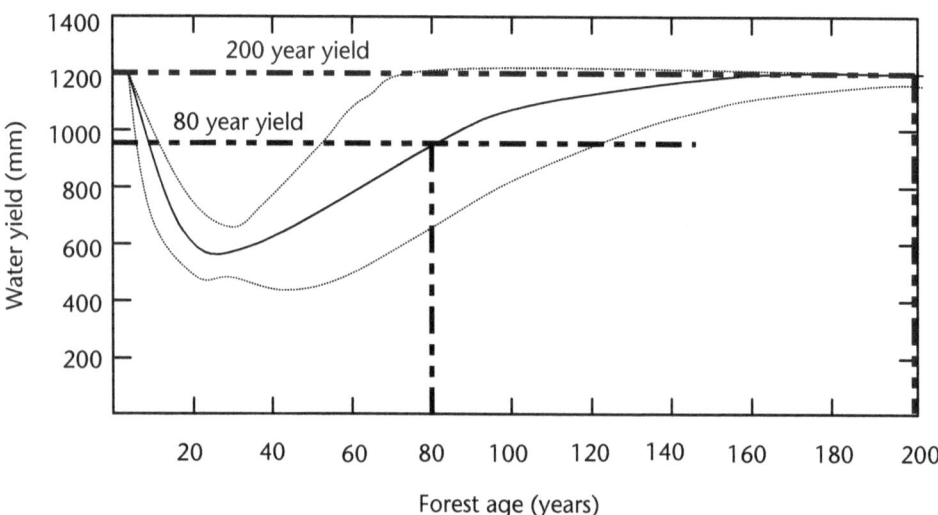

Figure 5.6. Effects of bushfire on water yield from mountain ash (*E. regnans*) forests (redrawn from Kuczera 1987).

We wrote in 2007 (Attiwill and Adams 2008):

'*Subsequent and continuing research (e.g. Roberts et al., 2001; Cornish and Vertessy, 2001; Lane and McKay, 2001) expanded the Kuczera generalisation of catchment behaviour following fire to E. sieberi, E. saligna and E. fastigata/E. cypellocarpa/E. obliqua in the east of Australia and to E. marginata/E. diversicolor/E. calophylla mixtures in the west. While there is some uncertainty about the magnitude of the responses (both the increase in yield in the first few years after fire, and the reductions in yield over longer periods) in forests other than E. regnans, the general pattern seems similar in all studies.*'

In support of this statement, we note that land and water management agencies in Western Australia have begun serious trials (the Wungong Catchment Trial) of silvicultural options for active forest management, including thinning and fuel-reduction fires that should increase water yield. This seems a highly responsible approach and one that could, or even should, be copied elsewhere. Encouragingly, early results are confirming that old jarrah (*Eucalyptus marginata*) forests use less water than young jarrah forests in Western Australia (Macfarlane *et al.* 2010) and that management can be used to create 'old-forest like' attributes (including greater water yield) in younger stands.

We caution here that while the general pattern of reduced water yield after bushfire does hold, not all eucalypt catchments, or at least sub-catchments, behave in identical fashion. Much depends on patterns of leaf area, understorey development and species-specific differences in anatomy and physiology that can also be site-dependent (for example, there are significant differences in understorey development depending on topography and aspect). At the extreme, for some steeply sloping, south-facing sub-catchments dominated by *E. regnans*, stable isotope analysis shows that incoming rainwater interacts only weakly with the vegetation, instead passing through to base-flow with minimal residence times (Pfautsch *et al.* 2010a). Although such forests are also the least likely to burn, at some time they will and the effects of bushfires there are likely to differ strongly to north-facing sub-catchments or on more gently sloping terrain.

Leaf area, plant water use and yield

Improving knowledge of the physiology of eucalypts and understorey plants is helping to refine our ability to predict the effects of fire on water yield. In large part, this is because the plant water use, or the *evapotranspiration*, component of the water balance for a given forest stand or catchment usually dominates the overall water balance.

Usually, the water yield from forested catchments is considered (and predictions made) using water balance equations of the form:

$$P = ET + R + D + \Delta S$$

where P is precipitation, ET is evapotranspiration (or plant water use), R is runoff (stream-flow), D is groundwater recharge (not contributing to runoff), and D S is the change in soil water storage. The precipitation term is sometimes modified to account for the amounts of water retained in canopies and quickly lost back to the atmosphere as leaves dry. Irrespective, the ET term dominates the water balance for most eucalypt forests in southern Australia, with as much as 80% of incoming water used by the vegetation, leaving relatively little for runoff.

The dominance is readily appreciated if the difference between actual rainfall and actual runoff (Observed E) is compared with the difference between actual rainfall and runoff estimated based on potential ET (Predicted E, Figure 5.7). For more than 100 catchments around

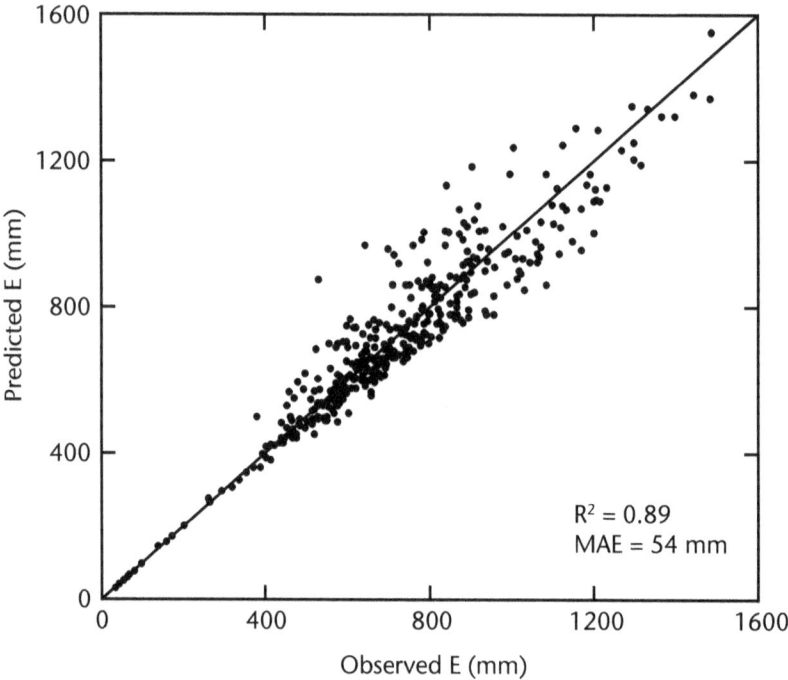

Figure 5.7. Scatter plot of observed and calculated evapotranspiration (E) for catchments around the world, including eucalypt-dominated catchments (after Zhang et al. 2004). MAE = mean absolute error.

the world, including a number of eucalypt forests (Zhang *et al.* 2004), the relation only deviates from the 1:1 line at higher rainfall.

Whitehead and Beadle (2004) provide a significant compilation of leaf area and evapotranspiration data for eucalypts, albeit a dataset dominated by plantations. Their conclusion, that: 'in well-watered conditions, transpiration from *Eucalyptus* forests can be explained largely by leaf area index and D' (where D stands for air saturation deficit), is moderately well supported by studies in forests.

In general, the pattern of water yield is directly related to total leaf area (including understorey) in southern eucalypt forests. For example, Wood *et al.* (2008) showed that water use by understorey in old-growth mountain ash forests could contribute up to 45% of the total plant water use – a fraction similar to that estimated by Vertessy *et al.* (2001). Pfautsch *et al.* (2010b) showed that even in younger forests (70 years old), water use by understorey (mainly *Acacia* and *Pomaderris* spp.) was a highly significant component of the total. Although most studies of understorey water use have come from the tall-open forests such as mountain ash, the results are uniformly clear – the understorey uses significant amounts of water.

E. regnans and other ash eucalypts (e.g. *E. delegatensis* and *E. sieberi*) regenerate after fire from seed. In Western Australia, most of the *E. diversicolor* (karri) forests regenerate from seed. However, a great many more eucalypts recover after fire by sprouting. For these eucalypts, we have poor knowledge still of the hydrological effects of the clearly different leaf morphology and whole-tree architecture. These, plus the rate at which canopy leaf area re-develops after fire, and the maximum leaf area that re-develops, are key variables in determining vegetation water use.

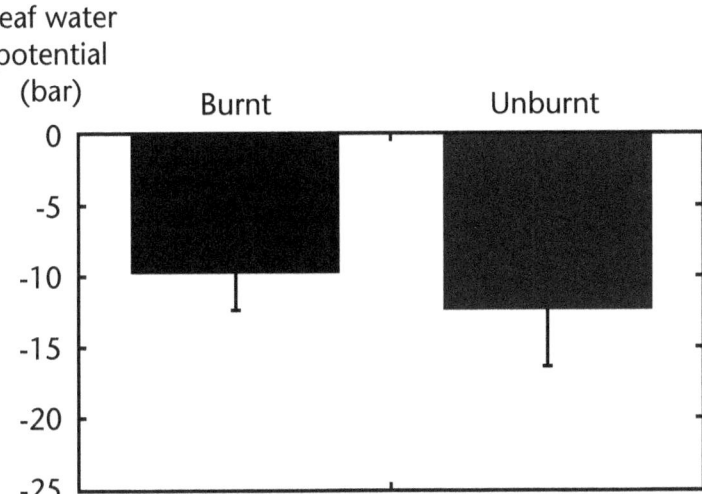

Figure 5.8.   Leaf water potential at dawn of snow gums (*E. pauciflora*) in areas treated with fuel-reduction fires in the Snowy Mountains (Turnbull and Adams, unpublished data).

Fuel-reduction fires and water availability

Despite the now clear importance of understorey to total stand water use, there has been little research on the hydrological effects of fuel-reduction fires. We recently initiated a long-term study of fuel-reduction fires in the Snowy Mountains. The study includes analysis of the inter-action between grazing and burning and is based on paired, adjacent plots of good size (0.5 hectares) that are either left 'as is' or that are burnt using standard fuel-reduction techniques, by the local Rural Fire Service. Some 'before and after' photos are shown in Colour plate 6.

A simple analysis of the hydrological effects of the fires is given in Figure 5.8. Pre-dawn water potential[18] in plants is a reasonably robust measure of the water potential in the underly-ing soils, given that plants have had all night to re-equilibrate with the soil and that night-time transpiration is a small fraction of day-time transpiration. We measured the mean water potential of the five species that contributed the most leaf area, including the overstorey *E. pauciflora*, over three consecutive days. Water potential in the unburnt plots was significantly less (more negative) than in the burnt plots. In other words, there was less water available in the unburnt plots – less water for plants, less water to become either runoff or groundwater recharge.

Summary – fire and water

The interaction between water and fire is an important subject in Australia because: (a) water is just about the most limiting resource we have – for people to drink, for industry and for agriculture; (b) bushfires greatly reduce (for decades) water yield from forested catchments, including those that supply the major cities of Perth, Adelaide, Melbourne, Canberra and Sydney, and those that provide most of the water in the Murray–Darling system; (c) we can mitigate the risk and increase the security of that yield through fuel-reduction fires that provide the additional direct benefit of increasing water yield.

As a recent example, Richard Benyon and others from CSIRO (2007) estimated the likely impacts of the 2003 fires for water yield to the Murray–Darling system. They focused on alpine ash (*E. delegatensis*) and snow gum (*E. pauciflora*) and wrote:

'However, as a result of increasing transpiration from the regenerating overstorey of the alpine ash and snow gum forest areas, by about 2015, mean annual transpiration in the Alpine (Catchment Management Unit) is predicted to increase by about 82 GL above pre-fire transpiration. Again, all other factors remaining unchanged, increased transpiration is expected to result in a maximum 130 mm reduction in mean annual stream flow from the catchment.'

Putting this into context, in 2008 water in the Murray–Darling was trading at more than $2000/ML. As Australian taxpayers know, hundreds of millions of dollars are being used to purchase water in the Murray–Darling basin for the environment. The loss due to bushfire of 82 GL[19] *per year* represents a real cost of more than $160 million each and every year to the Australian taxpayer, for several decades. It makes good sense to transfer a portion of that value to fuel reduction. Indeed, it seems obvious that the federal government could do far more for the environments in the basin by spending taxpayer funds on fuel reduction in the forested, national park catchments than by direct purchase of water from productive users of that water downstream.

Concluding remarks – carbon, water and nutrients

Properly conducted, fuel-reduction burning offers land managers the potential to achieve a positive environmental 'triple bottom line' – more available water and water yield in catchments, more carbon sequestered, and an improved N:P balance that will support herbaceous species, as well as the less P-demanding woody shrubs. The alternative of less fire in the landscape will mean far less security of water yield, a high probability that carbon emissions during bushfires will rise steeply and become a major contributor to Australia's total carbon emissions, and continuation of dominance of understoreys by sclerophyll shrubs with ecosystem 'collapse' becoming a significant possibility.

In heavily forested states such as New South Wales and Victoria, the carbon cycle of native forests will always be of far greater significance than tree planting. Native forests will be carbon sinks when young and carbon sources when old. We should prevent as far as possible, high-intensity, long-duration fires that consume long-stored soil carbon as well as that in plants.

Regarding fuel-reduction fires and carbon, Australian policies will almost certainly have to fall into line with those elsewhere. For example, in the USA (Stephens *et al.* 2009):

'Policies have been enacted to encourage carbon sequestration through afforestation, reforestation, and other silvicultural practises; however the effects of wildfires on forest C stocks are poorly understood'. Furthermore: 'Forest managers face an important decision: should C stored on site be maximised ... or should some of that C be removed using active treatments, including prescribed fire ... thereby reducing total stored C in the short term but increasing fire resistance in the long term?' And, finally: 'future work should investigate how C stocks change over time in response to fire hazard reduction treatments'.

6

Fire and climate change

Introduction

Changes in climate have been part of the Australian continent, and its flora and fauna, for millennia (see Hill 2004). As far as can be discerned from fossil and other evidence, the evolutionary history of the continent includes periods that have been wetter than at present and those that have been drier. These cycles of wetting and drying have corresponded in part with northward movement of the continent. Again, as far as can be determined, fires have increased and decreased in frequency, and likely in severity, alongside changes in rainfall and mean annual temperatures. These powerful, macro-scale, forces (climate and fire) have played major roles in the evolution of the flora and fauna over time scales measured usually in the tens of thousands to millions of years.

The patterns are reflected internationally. Bowman *et al.* (2009) summarised the role of fire in the Earth system and re-emphasised its critical nature to nearly every part of the globe. Their analysis (Figure 6.1) includes an assumption of a greater role of fire in the future.

Returning to Australia, the present day flora is interpreted through a 'fire lens' by nearly all biologists. For example, the dominance of sclerophyll (hard leaves) vegetation over much of the southern half of the continent is widely described as being due to the relative aridity of

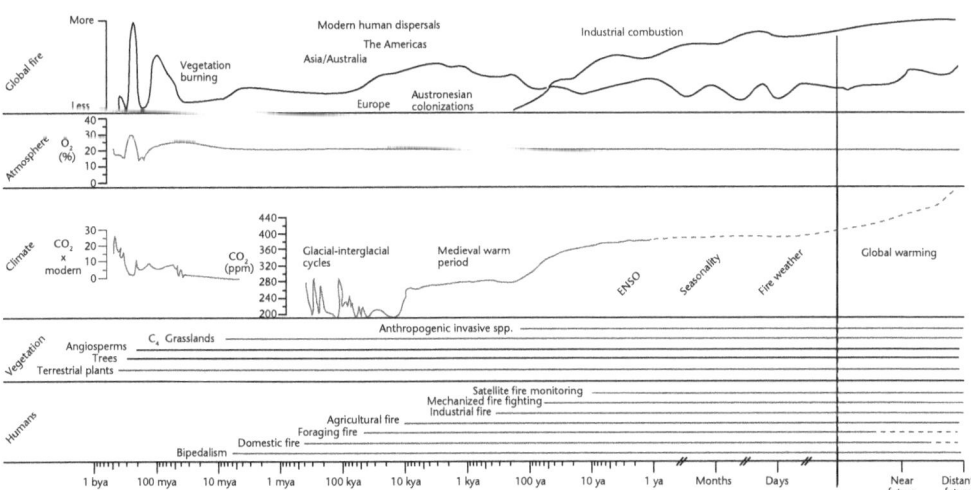

Figure 6.1. Schematic of global fire activity in relation to atmospheric CO_2 and O_2, appearance of certain vegetation types and the presence of the genus *Homo* (from Bowman *et al.* 2009).

more recent millennia compared with previous wetter periods. The dominant sclerophyll vegetation is also supremely well adapted to fire – most species require fire at one frequency or another for their reproductive success. There are, however, even in landscapes dominated by sclerophylls, pockets of non-sclerophyll vegetation that have persisted where the landforms have protected them from the more generally arid conditions (e.g. Adams 1996).

Classic examples in Victoria and Tasmania are the relict cool-temperate rainforests dominated by *Nothofagus* in deep, well-sheltered (e.g. south-facing) valleys in the wettest parts of landscapes that are otherwise dominated by sclerophyllous *Eucalyptus* species. *Nothofagus* and native conifers (e.g. *Podocarpus* spp.) that take long periods to regrow after fire become more dominant in the cooler and wetter climates of western Tasmania and at higher elevations in Victoria and Tasmania. Even so, much of the higher country in Victoria and Tasmania is dominated by grasses and sclerophyll shrubs that are clearly fire-adapted and even fire-dependent. It would be wrong to suggest that the presence of the relictual species in any general area means that area has not been burnt or should not be burnt. The topography–weather–climate interaction that allowed their persistence in a sea of sclerophylls will either continue to protect them or, as climatic conditions change, it won't. Attempting to prolong the persistence of temperate rainforest species in places where climates and consequent fire regimes are no longer suitable is not an attempt at conservation – it is an attempt at preservation and, in all likelihood, will fail.

In the far south of Western Australia, where rainfall reaches a maximum and droughts are rare, there are again numerous examples of plant species that recover only slowly after fire and that may require long periods without fire to resume their present day abundance (e.g. Wardell-Johnson and Burrows 2004). The sandstone escarpments north and south of Sydney on the east coast, provide another clear distinction between the sclerophyll vegetation that dominates the tops of escarpments and ridges on the one hand, and the often more mesophytic vegetation that is found in the valleys. The classic example is undoubtedly the Wollemi pine. The valleys may require a different fire regime to the ridges if preserving current patterns of abundance and dominance is regarded as desirable. Again, we emphasise that if the climate changes sufficiently, attempting preservation by excluding fire is likely to be difficult, at best.

Within these broad, landscape-scale patterns lies a myriad of sub-patterns. The most well known are probably the changes with time in abundance of different functional groups. The often slower growing 'sprouter' species require longer periods without fire to become significant components of the understorey of many forests while 'seeder' species can be eliminated by too many fires in too short a period (see Bradstock *et al.* 2002).

Thus, at the landscape scale, forests and most other vegetation types are mosaics, with species composition varying according to landscape and fire history. Indeed, most do not have a constant species composition either at any particular geographic point, or at any point in time. Instead, the species composition of most ecosystems is in a state of flux depending on the time elapsed since the last fire.

As a consequence of the importance of fires in determining the species composition of nearly all ecosystems in southern Australia, it is arguable that changes in climate will modify biodiversity through effects on fires, far more strongly than direct effects of climate on individual species (Figure 6.2).

Future climates and fire weather

Recent reports prepared by CSIRO and the Bureau of Meteorology for the Climate Institute of Australia (e.g. Lucas *et al.* 2007), illustrate the increasing risks of fires due to changes in climate

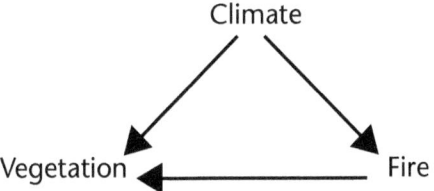

Figure 6.2.   Climate may act directly on vegetation, or indirectly via fire regime.

and weather. Cumulative forest fire danger indices (ΣFFDI) for Melbourne, Sydney and Adelaide, and for Cobar in inland New South Wales, were computed and are shown in Figure 6.3. One of the notable features has been the 'step' increase in risk in inland New South Wales. Lucas *et al.* (2007) attributed this as follows:

> *'This tendency is particularly pronounced in interior NSW since the turn of the century. The changes in ΣFFDI at these stations are associated with an increase in the number of VHE (very high – extreme) days.'*

In addition to the increase in FFDI over recent decades is an increase in the length of the 'fire season'. As shown in Figure 6.4, that too has increased over recent decades, ranging between 2 and 6 days extra each year among Melbourne, Adelaide, Canberra and Wagga.

Projections of climate are, at best, a difficult business. Nonetheless, we note that both Lucas *et al.* (2007) and Pitman *et al.* (2007), using different approaches and models, and two emissions scenarios, came up with similar conclusions. They are worth quoting:

> *'In general, fire weather conditions are expected to worsen. By 2020, the increase in ΣFFDI is generally 0–4% in the low scenarios and 0–10% in the high scenarios. By 2050, the increase in generally 0–8% (low) and 10–30% (high). The largest changes are expected in northern New South Wales. Little change is expected in Tasmania. With this increase in ΣFFDI, a larger number of days with a Fire Danger Rating of 'very high' or 'extreme' are*

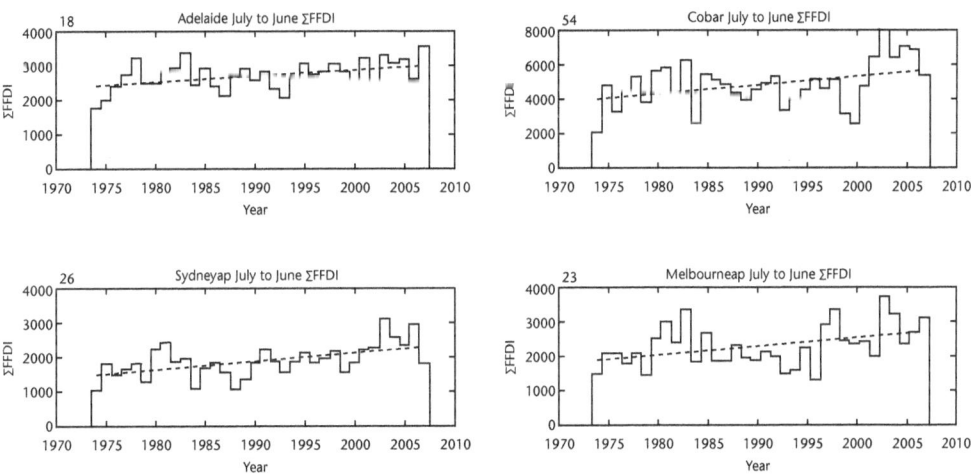

Figure 6.3.   Cumulative Forest Fire Danger Indices over the past ~35 years (from Lucas *et al.* 2007) – ap after Sydney and Melbourne refers to airport.

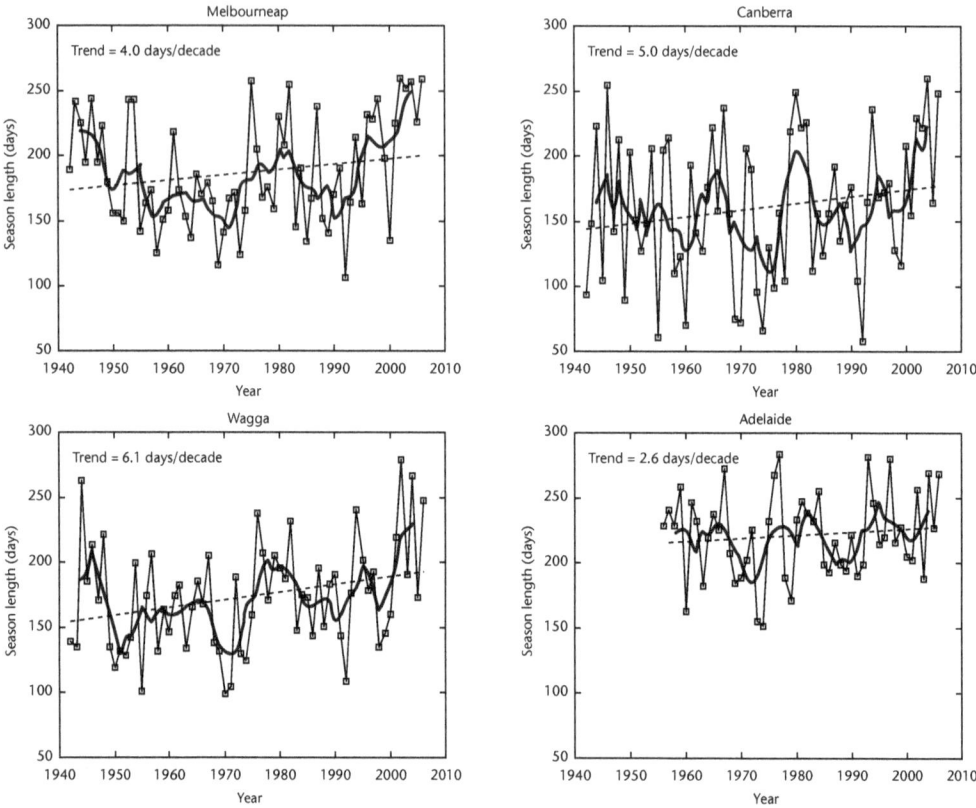

Figure 6.4.   Length of fire season at four sites in south-east Australia over past 40–60 years (from Lucas *et al.* 2007) – ap after Melbourne refers to airport.

also expected. The number of 'extreme' fire danger days generally increases 5–25% for the low scenarios and 15–65% for the high scenarios. By 2050, the increases are generally 10–50% in the low scenarios and 100–300% for the high scenarios. The seasons are likely to become longer, starting earlier in the year.' (Lucas et al. 2007).

'Results show a consistent increase in regional-scale fire risk over Australia driven principally by warming and reductions in relative humidity in all simulations, under all emission scenarios and at all time periods. We calculate the probability density function for the fire risk for a single point in New South Wales and show that the probability of extreme fire risk increases by around 25% compared to the present day in 2050 under both relatively low and relatively high emissions, and that this increases by a further 20% under the relatively low emission scenario by 2100. The increase in the probability of extreme fire risk increases dramatically under the high emission scenario by 2100.' (Pitman et al. 2007).

Changes in climate, forest productivity and fuels

Although it is probably self-evident that a drier and hotter climate would increase the risk of bushfire through drier fuel and a greater number of days when climatic conditions are most conducive to fire (high temperatures, low relative humidity and strong winds), there are

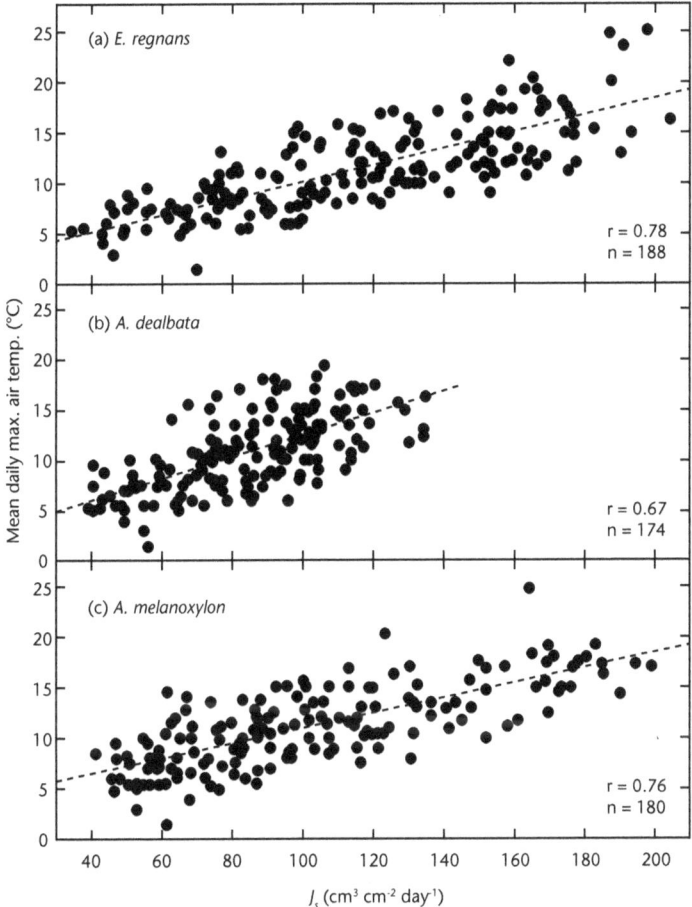

Figure 6.5.   Relation between maximum daily air temperature and sap flow (J_s or evapotranspiration, see also *ET* in Chapter 5) for mountain ash (*E. regnans*) forest near Warburton, Victoria (after Pfautsch *et al.* 2010b).

mitigating factors. One of the most important will be possible changes in productivity and water balance of forests and other vegetation.

Net primary productivity (NPP)[20] is the major determinant of fuel loads, especially the mass of fine fuels such as recently fallen leaf litter, bark and wood, and the mass of the understorey. Likewise, the water balance of a forest stand (most strongly regulated by rainfall and tree water use) will determine the moisture content of the litter and understorey (e.g. Matthews 2006). As a consequence, changes in productivity and water balance will affect fire risk. Here there is considerable uncertainty as to the response of southern Australian forests to changing climates. In its simplest guise, an increase in temperature may mean greater water use by the vegetation in forests and thus less water yield. As shown in Figure 6.5, both overstorey (*E. regnans*) and understorey (*Acacia* spp.) in mountain ash forests transpire water at temperature-dependent rates. Temperature may be a proxy for intercepted radiation in this forest type, but its effect may also be due to the more constant humidity of the *E. regnans* landscapes relative to many others in Australia.

Any increase in temperature is only likely as an accompaniment to increased concentrations of CO_2. Hence, an increase in water use efficiency could offset the increase in water use

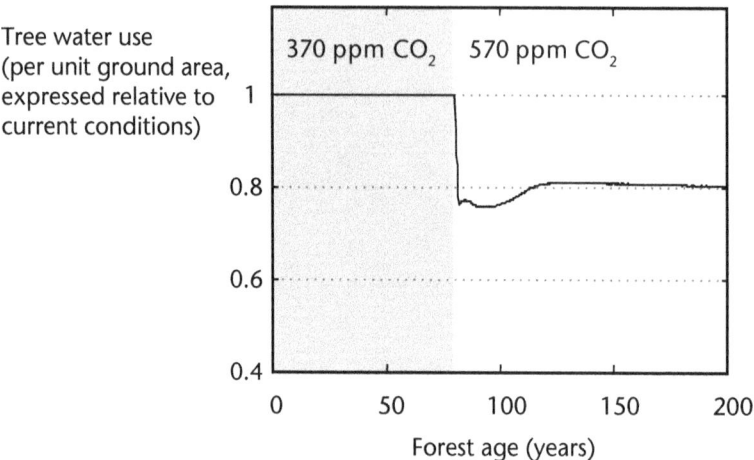

Figure 6.6. Predicted (see Buckley 2008) changes in evapotranspiration by eucalypts under current and future concentrations of atmospheric CO_2.

due to temperature. Such an increase is widely reported from glasshouse studies that also report increases in productivity due to the 'carbon dioxide (CO_2) fertilisation effect'. However, there are many other constraints to productivity, especially in Australia. The most significant is water, but that is followed closely by limited availability of major nutrients such as nitrogen and phosphorus.

Using the DESPOT model of forest growth (e.g. Buckley and Roberts 2006), recent Bushfire CRC research (Buckley 2008) suggested increases in the efficiency of growth (amount of carbon gained per unit of water used) may at least partly overcome water limitations (Figure 6.6) such that growth is unaffected while runoff may actually increase. We note here that this prediction about the overstorey and water is not at odds with earlier comments about fuel-reduction burning being capable of increasing water availability. In fact, enhanced growth under future climates will also increase fuel loads. The win–win for increased water yield and reduced risk of bushfire that could come from managing those loads seems all the more important if current climate predictions come true.

Predictions from the DESPOT model are strongly supported by data from a range of 'free air carbon enrichment' (FACE) studies that show increased NPP, along with increased leaf area, but a significant reduction in stomatal conductance and decreased water use (Ainsworth and Long 2005). This is further supported by the Intergovernmental Panel on Climate Change (IPCC), which suggested that the best available process-based models of productivity indicate that terrestrial ecosystems, especially forests, could take up between 22 and 57% of expected anthropogenic emissions (e.g. IPCC 2001).

In addition to modelling, a major collaborative research effort is underway in the Sydney basin to elucidate responses of eucalypts to high CO_2 (Colour plate 7). As shown in this photograph, the whole tree chambers provide a means of assessing the interactive effects of CO_2 and water on tree growth (and thus litter and fuel production) and water use (see also Barton *et al.* 2010). Such approaches are crucial if we are to assess the likely effects of changing climates and atmospheric CO_2 on fuel loads, and thus fire risk.

We have much poorer information as to how nutrient limitations might affect carbon and water outcomes – some scientists predict that nutrient limitations will quickly counter any CO_2 fertilisation effect, while others point to increases in nutrient availability that could result from enhanced growth and nutrient cycling. Hungate *et al.* (2003a) used known ratios of carbon to nitrogen in trees (~200), wood (~500) and soils (~15), and the same models as the IPCC, to calculate the amounts of 'extra' nitrogen that would be required to synthesise the 'extra' biomass and soil carbon. They calculated significant shortfalls between the amounts of nitrogen required and that likely to be available. Even allowing all the simulated increase in tree carbon to accumulate in wood only slightly reduced the amount of N required (owing to small absolute changes in amounts of N) and the shortfall.

Hungate *et al.* (2003a) established that failure to take nutrient cycling into account would likely result in an over-estimation of the capacity of terrestrial ecosystems to take up CO_2. However, this also depends on fire frequency (Hungate *et al.* 2003b). Although Hungate *et al.* (2003a) made reasonable provision for biological nitrogen fixation, Dan Binkley and colleagues (e.g. Binkley *et al.* 2000; Resh *et al.* 2002) have shown that introduction or invasions or even 'natural expansions' of nitrogen-fixing trees can produce 'massive changes in soil N cycling' (Binkley 2005). There remains a possibility that increases in atmospheric concentrations of CO_2, particularly if accompanied by changes to climate and fire regimes, may result in increased abundance of N-fixing species that could significantly increase soil carbon storage and lower soil C:N ratios. Binkley (2005) also noted that rates of cycling of P always seem to be greater under N-fixing species – thereby solving another problem of the large requirement for P by biological nitrogen fixation and possible future P-limitations (see also Chapter 5). Lloyd *et al.* (2001) modelled the availability of soil phosphorus under a high CO_2 environment for moist tropical forests. This modelling suggested that phosphorus availability may increase to match the increase in rates of carbon fixation. Barrett and Gifford (1995) and Barrett *et al.* (1998) had earlier shown that one of the adaptations of crop plants to a high CO_2 world could be an improved ability to mineralise soil organic P. Much of the speculation regarding nutrient availability under future climate and fire regimes will remain just that until much more research has been done.

Changing climates and plant adaptations

Changes in climate will have direct effects on plant growth, as will increased concentrations of atmospheric CO_2, but these will be at least partly accommodated by adaptations in plant structure and function. The historical record (e.g. fossil record) shows us that within species, the density of stomata (number per unit leaf area) changes with the atmospheric concentration of CO_2 – in other words (and given sufficient time), the potential increase in productivity from 'CO_2 fertilisation' could be offset by reduced density of stomata such that there is little net change in CO_2 concentrations inside leaves (e.g. Woodward 2002; Woodward and Kelly 1995). Likewise, there is ample evidence that plants adapt many other fundamental biochemical processes to changes in temperature or water availability such that they remain productive and reproductively viable. At present there is insufficient data to be confident about predictions of changes in fuel loads or fuel moisture content – and thus fire risks.

Nevertheless, it seems useful to consider what effects changes in fire frequency and severity (fire regimes) would have on plant diversity and abundance. Grasses and herbs are just as important biodiversity components as are the more obvious and more commonly discussed woody shrubs. What we can say is that the decline in number of fuel-reduction fires in states such as Victoria and New South Wales over the past 20 years (Figure 6.7) is associated with increasing abundance of woody shrubs.

A possible outcome of climate change (hotter, drier conditions) is an increase in frequency of lightning-started fires that are *not* suppressed. An alternative is that the use of fuel-reduction

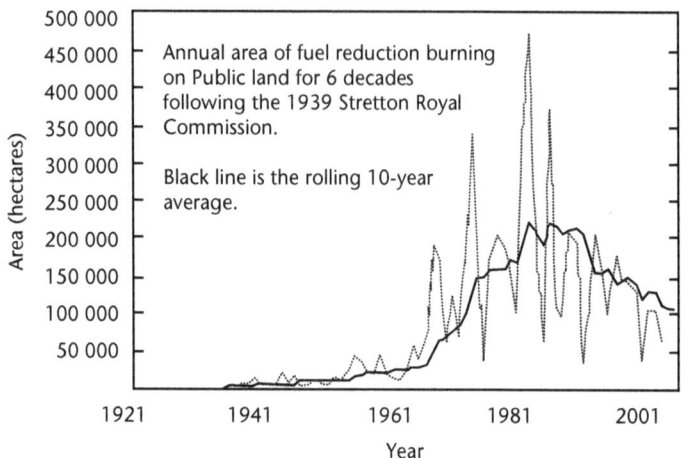

Figure 6.7. Annual area of fuel-reduction burning on public land in Victoria from 1940 to 2003 (after Tolhurst 2003).

fires is increased. In relation to the reversal of current trends of increasing incidence of woody shrubs, either would produce a similar outcome. Perhaps the most concerning outcome would be that there is no increase in deliberate use of fire, and all natural fires are extinguished before any significant area is burnt. This could lead to large increases in the area of land on which the fuel load of litter accumulates and the understorey shifts ever more strongly toward woody shrubs. The risk of large, uncontrollable bushfires would quickly increase to near certainty.

Changes in climate: fire and carbon budgets

The respiration of soil and litter (or heterotrophic respiration) is one of the most important processes in the global carbon cycle – a change in just 10% in soil organic carbon would be equivalent to all the anthropogenic CO_2 emitted over 30 years (Kirschbaum 2000). Losses of soil carbon via respiration are highly temperature sensitive. Although the same is true for plant respiration, the sheer amount of carbon stored in soils dictates a strong focus on soil respiration.

To illustrate the significance of soil carbon a little further, we can examine variation in soil respiration (the oxidation of soil organic matter by soil microbes) with temperature across the vegetation types (snow gum woodland and grassland) found in the high country (>1200 m above sea-level) of south-eastern Australia (Jenkins and Adams 2010). Under laboratory conditions, rates of respiration differed more than five-fold across communities that in the field are separated by little more than a few metres. Differences between communities became more pronounced as temperatures increased. Crucially, rates of respiration of soils from communities that differed only in understorey composition were markedly different. Other soil properties were equally spatially variable in this landscape, especially key variables such as phosphorus availability and the ratio of carbon to nitrogen (Figure 6.8).

The distribution of vegetation types and the abundance of shrubs in the understorey relative to that of grasses, is due to *both* micro-climate *and* fire regime. In addition to needing to know how fire will affect species composition from a biodiversity point of view, it will be at least as important to understand how fire regime affects crucial processes such as soil respiration through species composition. Already, there is considerable analysis and speculation in the ecological literature about the causes and consequences of changes in composition of

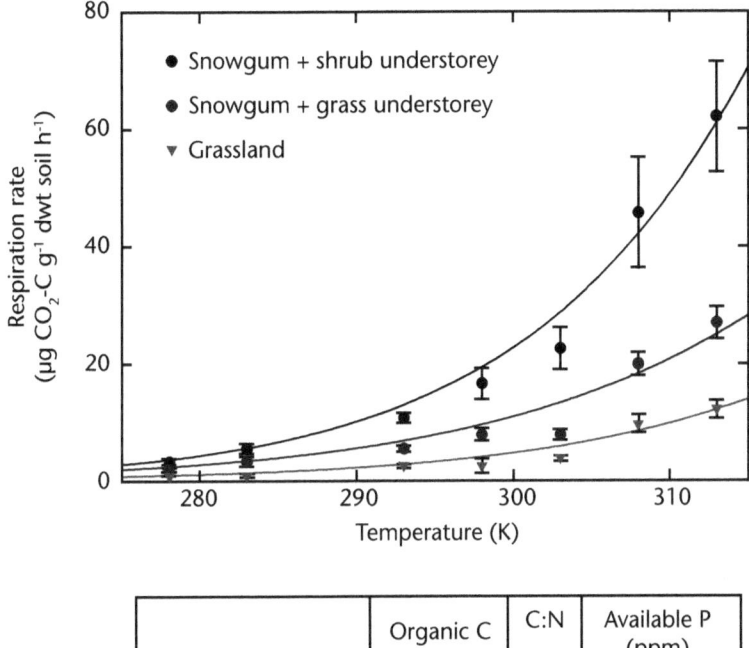

Figure 6.8. Rates of heterotrophic respiration by high country soils from the Snowy Mountains, New South Wales. Inset table shows chemical characteristics of same soils (after Jenkins and Adams 2009).

	Organic C (%)	C:N	Available P (ppm)
Snowgum + shrub	6.7	7.7	22.4
Snowgum + grass	6.3	10.7	11.1
Grassland	6.7	13.8	5.0

vegetation communities in the high country (e.g. McDougall 2003) and many other parts of Australia (e.g. Hughes 2003), and much research will be required to unravel the varying contributions of fire, grazing and climate to wholesale changes in vegetation before we can be sure of cause and effect (e.g. Fensham *et al.* 2005).

Irrespective of both proximal and distal causes of vegetation change, Australia's contributions to the global carbon budget will depend on the changes in soil carbon that accompany changes in vegetation, and how those then respond to changes in temperature and fire.

Summary

On geological time scales, Australian climates have changed dramatically. However, on scales as short as human lifetimes, events such as the recent drought in southern Australia might occur just once, rendering human interpretation of the effects of climate change, at best, a risky business.

That said, it seems that while many will focus on *potential* risks to biodiversity – either directly as a result of climate or indirectly via fire – the evolutionary history of the continent will ensure that we have the genetic 'foundation' for coping with changing climates and fire regimes. The sclerophylls will likely continue to dominate the mesophytes. One of the more immediate problems will be how we manage to maintain the more herbaceous but still sclero-phyllous species, and managed fire regimes can begin to take this into account.

Water and carbon are rapidly becoming among the most commonly used words in public life in Australia. Fires and fire regimes will play extraordinarily important roles in both – far more than the *direct* effects of climate change over the next 100 years or so. The importance of improved knowledge of the effects of fire on carbon (gains by, and losses from, soils) and water (especially catchment water yield) seems completely self-evident.

7

Fighting fire with fire: I. Why fuel-reduction burning? Does it achieve its aims?

Fuel-reduction burning is the planned use of fire to reduce fuels with the aim of reducing the intensity and spread of bushfires in subsequent years. Fuel-reduction burning is but one form of 'prescribed burning' – the lighting of fires under prescribed conditions to achieve specified management goals.

> 'Prescribed burning is the controlled application of fire under specified environmental conditions to a predetermined area and at a time and intensity required to meet specific management objectives. Some examples of prescribed burning include the use of low intensity fires to reduce fuel hazard levels to meet fire protection objectives, the use of high or low intensity fire to modify or renew habitats for specific faunal species, or high intensity fire to regenerate forests after timber harvesting, The range of possible management objectives requiring fire management is extensive' (Tolhurst 2003).

And another definition from the United States Department of Agriculture, Forest Service Southern Region (1989): 'Prescribed burning is fire applied in a skilful manner, under exacting weather conditions, in a definite place, and to achieve specific results'.

Prescribed burning, as defined above, sounds benign enough. However, the intentional use of fire arouses passions and conflicts. The forest manager faces an almost insurmountable problem: people, most of whom live in the cities, have an innate fear of fire. If fire is always seen as bad, how can it be used for good?

And, indeed, there is always a danger in using fire – a prescribed burn intended for 'good' becomes 'bad' when it escapes. We talk of the 'controlled application of fire', as in the definition above (and the term 'controlled burning' is often used synonymously with 'prescribed burning' – another term that is used is 'ecological burning'), but the control cannot always be precise.

Another term that could be used is 'planned fire'. Athol Hodgson, a former Commissioner for Forests and Chief Fire Officer in Victoria, writes (2004) that a key element of a policy for managing forest fires must be that the forest manager must:

> 'differentiate between unplanned fires that must be controlled, and planned fires that are lit to achieve specified outcomes. The former include natural fires caused by lightning that occur at times when they threaten or may threaten biodiversity and other assets on forests and lives and assets on adjoining lands. Planned fires include those lit to reduce fuel loads that increased as a consequence of controlling natural fires and fires lit to maintain the health and biodiversity of forest ecosystems or for regeneration of disturbed or degraded ecosystems.'

Hodgson goes on to say:

'A mind set within sections of the community that sees planned or 'prescribed' fire solely as a tool for production forestry or for protecting assets on adjoining private land is out of step with current thinking of mainstream ecologists and environmentalists. They (ecologists and environmentalists) know that fire is necessary for the health and biodiversity of most forest ecosystems and prescribed burning is an ecologically-conscious tool to achieve that objective and to mitigate the damaging effects of large intense wildfires.'

We will concentrate here on prescribed burning to reduce fuel loads in the bush, and this is often called simply 'fuel-reduction burning'.

During a discussion about the effectiveness of fuel-reduction burning, one of our colleagues stated forcefully that of course fuel reduction is effective. 'If there is no fuel, then there can be no fire', he said – 'there are no bushfires in the Sahara Desert'. Likewise, 'No Fuel No Fire' is the simple and strong slogan of the Victorian Lands Alliance. At the start of every fire season, the fire agencies urge landowners to clean up their properties – reduce fuel loads and thereby reduce the risks of damage from bushfires.

'No Fuel No Fire' is indisputable. Fuel, oxygen and heat are the elements of the fire triangle; remove any one of them and there can be no fire. This is straightforward and simple, and we all know these essentials when dealing with a campfire. We light the fire with a match, and encourage the fire to start with kindling. If we want more heat, we add more wood, and if we want to put the fire out, we remove the heat by pouring water on the fire or we remove oxygen by smothering the fire with dirt or we remove the fuel. Similarly, if we can get to a bushfire early enough, we can control the fire by removing the fuel with hand tools or bulldozers, and by removing the heat by spraying water directly onto the fire.

Testing the hypothesis: 'Fuel-reduction burning decreases intensity and rate of spread of subsequent bushfires'

The hypothesis is proven in theory

The intensity of a fire is calculated, as we discussed previously in Chapter 3, from the Byram Fire Intensity Index (I):

$$I = HwR$$

where H is the heat of combustion (joules per gram of fuel), w is the mass of available fuel (grams per square metre) and R is the rate of spread of the fire (metres per second). Intensity, the product of these three variables therefore has the units joules per second per metre of fire front, and since 1 watt = I joule per second, intensity (I) is given as watts per metre (W m^{-1}).

H, the heat of combustion of the fuel, is more or less constant: a value of 18.6 kilojoules per gram is applicable to most forest fuels. A difficulty of the Byram Fire Intensity Index is that the two variables w (the weight of available fuel) and R (rate of spread of the fire) are not independent. It is obvious that, for a given set of conditions, the greater the fuel load, the faster the fire will spread. In fact, Alan McArthur long ago noted a direct proportional relationship between these two variables under conditions of relatively mild weather. That is, if fuel load is halved, the rate of spread is also halved. Although some (Fernandes and Botelho 2003) claim that this direct relationship is now well substantiated, there is little supporting statistical evidence. However, it has been used in the literature for a long time (for example, by Hodgson in 1968) and it can be accepted as an approximation until such time that rigorous experiment or observation suggest otherwise (McCaw *et al.* 2008; Packham and Attiwill 2009). We can therefore say that the Byram Fire Intensity Index is proportional to the weight of available fuel squared:

$$I \propto w^2$$

Thus, if we reduce the weight of available fuel to 50%, the Fire Intensity Index is reduced to 25% (and, again, we stress the approximation). And, obviously, if there is no available fuel ($w = 0$), then $I = 0$, no matter what the conditions. Quite simply, the hypothesis is obvious, both as commonsense and in theory.

Is the hypothesis proven in practice?

In a bushfire, fuel loads and other forces such as weather and topography that promote the spread and intensity of the fire – both *in situ* and by carrying embers and fire brands well ahead of the main fire – are variable and, to a major degree, uncertain.

First, what do we mean by 'available fuel'? Fuel-reduction burns are generally done in the cooler months of autumn or spring when the fire can be comfortably controlled. These cool burns only consume a small proportion of total fuel that may burn under the most extreme weather conditions. They may burn only a small fraction of the so-called 'elevated fuels' – shrubs, the dry fibrous bark of the stringybarks, the long ribbons of bark hanging down from gums and ashes – that take the surface fire up into the crowns and start a crown fire. In contrast, a bushfire burning on a day of high danger can burn a vastly greater proportion of total fuel, leaving only the tree stems and larger branches above ground.

Thus we have an unnerving situation. On the one hand, it is practical commonsense to say that the removal of some of the fuels will result in the specified goal of reducing the spread and intensity of subsequent bushfires. On the other hand, it is often difficult to demonstrate quantitatively and scientifically that the specified aim of fuel-reduction burning has been achieved. Tolhurst and Cheney (1999) put these difficulties of assessing available fuels (and of determining the effectiveness of fuel-reduction burning on subsequent fire behaviour) into perspective:

> 'Because the available fuel depends on how the fuel is arranged, the distribution of moisture within the fuel bed and the intensity of fire created by the prevailing weather, it cannot be measured in advance of the fire. It has to be estimated from current and antecedent weather and presents a major problem in quantifying fuels for fire behaviour predictions'.

In other words, 'the weight of available fuel' is not a fixed quantity, but a quantity that varies among other things with size of the fuel, the three-dimensional distribution of fuels, the moisture content of fuels (a function of previous weather), and the weather prevailing at the time of the fire.

Consider an area that is planned for fuel-reduction burning. Available fuel before the burn was assessed as 10 tonnes per hectare. The burn is done on a perfect day in autumn, with a rate of spread of 100 metres per hour, and that all of the available fuel (as assessed) is burnt. The line intensity is then:

$$I = (18\ 600 \times 1000 \times 100/3600)\ \mathrm{Wm^{-1}} = 0.5\ \mathrm{MWm^{-1}}$$

and a good crew with hand-tools can readily control a fire of that intensity.

And so, as practitioners in the field, we would be told in the common parlance that 'This area has been fuel-reduction burnt'. What does this really mean in practical terms of fire prevention? For example, if the weather was not so good for fuel-reduction burning, and if there had been recent rain, then we could have the situation where (say) only 50% of the area was burnt. Thus we still have 50% of the area with a fuel load of 10 tonnes per hectare. Given a fire day with a rate of spread of 500 metres per hour, this 50% of the area would burn with an intensity of 2.5 megawatts per metre, and that intensity is close to the limits of suppression by any means.

That is only one part of the complexity. To quote from McCaw *et al.* (2008):

> '*Systematic analysis of interaction between bushfire behaviour and such factors as vegetation type, weather conditions and fuel characteristics as modified by fuel reduction burning has never been possible due to the unplanned and chaotic nature of bushfire events. There are always uncertainties about the fuel load on areas burnt by fuel reduction at the time of the bushfire which result from uncertainties about the areas actually burnt during the operation, the fraction of fuel removed and the subsequent rate of fuel accumulation. Accurate measurements of rates of spread and wind speeds during the bushfire are almost unattainable on comparable sites*'. *(Note: we have replaced 'wildfire' and 'prescribed burn' in this quote with 'bushfire' and 'fuel reduction burn'.)*

The point we are making is that the answer (yes or no) to the seemingly simple question – 'Was this area burned for fuel reduction?' – may or may not provide useful information if we are to assess the effect of fuel reduction in subsequent fire behaviour. There is a great deal of anecdotal evidence on both sides of the argument about the efficacy of fuel reduction, but anecdotal evidence cannot be taken further without further knowledge and at least some quantitative analysis.

Methodology to test the hypothesis in practice
Simple observation
There are many records of the effects of a previous fire on subsequent fire. For example, photographs taken following the 1961 Dwellingup bushfire (McArthur 1962) showing reduced fire intensity in areas burned 2 years previously are a valuable historical record.

Similarly, an area burnt in 1978–79 near Cann River, Victoria was largely unburned by a subsequent fire in 1983, despite severe weather conditions (Rawson *et al.* 1985). This sort of evidence is difficult to assess quantitatively because the amount of fuel reduction in the previous bushfire may have been much greater than that in a fuel-reduction burn. Nevertheless, it is a simple illustration of the effect of low fuel quantity in halting the spread of a bushfire, even under severe weather.

Underwood *et al.* (1985) used a slightly different approach. They analysed nine fires in the south-west of Western Australia from 1969 to 1984, and they used fire behaviour projections to compare actual and projected fire size. In every case, these comparisons indicated that previous fuel reduction reduced the size of the fire and improved ease of control.

A specific example of the effectiveness of fuel reduction comes from Landsat imagery of a forest fire burning under severe conditions not far from Perth, Western Australia, January 2007 (Colour plate 8). There were many similar examples coming from the 2002–2003 bushfire season in the southern half of Western Australia; that season was the most severe in 42 years as a result of at least 5 years of drought and severe lightning storms. A total of 549 bushfires covered about 126 000 hectares of Department of Environment and Conservation lands in the south-west. This was an eight-fold increase on the annual average over past years. Of all of these fires, 87% of all fires were kept to less than 100 hectares, mainly due to the presence of low fuels that had been the result of fuel-reduction burns in the past 6 to 8 years. All of the largest fires were contained once they ran into fuel-reduced areas. This includes the Mt Cooke fire (18 000 hectares) located 60 kilometres south-east of Perth. This fire was confined by five large prescribed burns ranging in age from 1 to 5 years old. Similarly a bushfire in the Walpole Wilderness burned 29 000 hectares, the largest forest fire since 1961; it was eventually contained within relatively low fuels within forested blocks that had fuel-reduction burns in the past 6 years.

The most-quoted review of the effectiveness of fuel-reduction burning in subsequent fire behaviour is that of Fernandes and Botelho (2003). They cite many well-documented examples that 'testify to both the virtues and limitations of hazard-reduction burning'. Studies from Western Australia feature prominently among those showing the virtues of fuel reduction, even after 10 years in *Eucalyptus* heathlands, and even in weather conditions that generated fireline intensities as high as 20–40 megawatts per metre. An analysis of the fires in Victoria over the period 1990–1997 showed that fuel reduction was effective in only 11% of the fires, this effect being strongest where fuel reduction was within 2–4 years of the fire (McCarthy and Tolhurst 2001); it should be noted, however, that fuel reduction in Victoria was only of the order of 200 000 hectares during the 1980s and fell further to 150 000 hectares during the 1990s.

The prolonged alpine fire of 2002–2003 in north-eastern Victoria burnt more than 1 million hectares, and the crowns of the trees were completely burnt or scorched over some 50% of this area. It is clear from aerial photos that a number of recent fuel-reduction burns were either not burnt, or only lightly burnt, by the bushfire. A quantitative analysis of this effect of fuel reduction has not yet been published.

Many studies reviewed by Fernandes and Botelho (2003), both in Australia and overseas, found that fuel-reduction burning has significant effects in subsequent fire-fighting and in protection of assets. Fernandes and Botelho conclude:

'Quantification of the influence of fuel reduction on large bushfires remains elusive, but the existing evidence supports the conclusion that recently treated areas do limit the spread of a fire and will result in a less homogeneous post-burn landscape. It is clear that fuel-reduction fire moderates bushfire severity and can benefit bushfire control operations in various ways, by increasing the safety of the personnel involved in suppression, decreasing the quantity and type of fire fighting resources (e.g. ground crews instead of aircraft), changing the overall suppression strategy (e.g. direct attack instead of indirect attack), reducing the risk inherent to the burning-out operations that are used in indirect attack, lessening the amount of mopping up, or simply providing better access and anchor points for suppression actions'. (Note: we have replaced 'wildfire' and 'prescribed burn' in this quote with 'bushfire' and 'fuel reduction burn'.)

These are benefits indeed, but if we are to be quantitative in our assessment – particularly when we are dealing with large, intense fires – then we have to go further than observation.

Analysis of past events

For many fire ecologists and fire researchers in Australia, the fuel-reduction program in the forests of south-western Western Australia has set the benchmark in how it should be done and what can be achieved.

Before 1961, fuel reduction in Western Australia was on a relatively small scale, averaging about 4000 hectares annually. On 19 January 1961, however, devastating fires started around Dwellingup and, following severe conditions a month later, much of south-western Western Australia was ablaze. The Dwellingup fire of 1961 was so serious that the State Government ordered a Royal Commission to investigate the circumstances and management of the fire. Emergency Management Australia[21] summarised the fires:

'On 19 January, after several days of hot weather and low humidity spawned several fires, a chain of thunderstorms swept the area extending from Mundaring in the north to Manjimup in the south. Between 5.30 and 6 pm, 10 fires started from lightning strikes 20–25 km from Dwellingup, 110 km south of Perth.

On 20 January, thunderstorms again swept the area, starting 9 more bushfires in State forests near Dwellingup. In addition, fires were burning at Gooseberry Hill, Kalamundah, Glen Forest, Mellan, Dandalup, Greenmount, Denmark and Kingsbrook. Over 40 000 hectares had been burnt in 4 days.

After 4 weeks of constant firefighting, on 24 February, the temperature rose to 41⁰C and relative humidity dropped to 14%. Because of a tropical cyclone to the north, strong winds of 50–60 km per hour blew into the south-west of the State, igniting all the extensive forests. Many small timber towns deep in the forests were hastily evacuated. Dwellingup became an evacuation centre, as police at Pinjarra, 24 km away, received a message that fire, completely out of control, was approaching the town, containing 1000 people. The last message to Pinjarra Post Office from Dwellingup reported that the local garage had exploded, and that houses on both sides of the Post Office were in flames. All roads into the town were cut by fire. Ionisation of the atmosphere by the fire blacked out communication to the Forestry Department's portable two-way radio in the town. When the fire had passed and the radio became operable again, it was learnt that no serious casualties had occurred. All the people had been moved to a large open space near the town's centre and escaped the fire. However, 132 homes, 2 service stations and 3 shops had been destroyed. Nearby timber villages of Holyoak and Nanga Brook were almost completely destroyed, with no casualties, as residents were evacuated to Dwellingup. Between 11–15 February, blow-up conditions again occurred and moderate-heavy damage caused in the Manjimup-Pemberton-Shannon River area. In early March, serious fires hit the Augusta-Margaret River area. From January to March of that year, about 1.8 million hectares had been burnt in the south-west, with consequential large loss of livestock'.

The Forests Department of Western Australia responded immediately to these fires and to the Royal Commission by major improvements in fire-fighting capability, including equipment, radio communications and weather forecasting. The Department also set up a world-class research unit in fire behaviour, and started an improved and greatly enlarged program of fuel-reduction burning.

From 1961 to 1990, annual fuel-reduction burning averaged about 300 000 hectares, and from 1990 to 2005, about 150 000 hectares. In the last few years, annual fuel-reduction burning has been in the range 6–8% of forested crown lands now managed by the Department of Environment and Conservation, Western Australia, with the current aim being 200 000 hectares annually (8%). The intensity of this fuel-reduction program can be judged by comparison with fuel-reduction programs in the other Australian states of 2% or less annually. For example, in the year 2008–2009, the total area of fuel-reduction burning in Victoria was 151 000 hectares, or 2% of the forest area managed by the state.

Rick Sneeuwjagt, Department of Environment and Conservation (DEC), Western Australia, analysed the data on prescribed burns and bushfires in south-western Western Australia over the 46 years 1961–62 (following the Dwellingup fires) to 2007–08 (Sneeuwjagt 2008). He calculated the average area burnt by fuel reduction for each 4-year period, and the average area

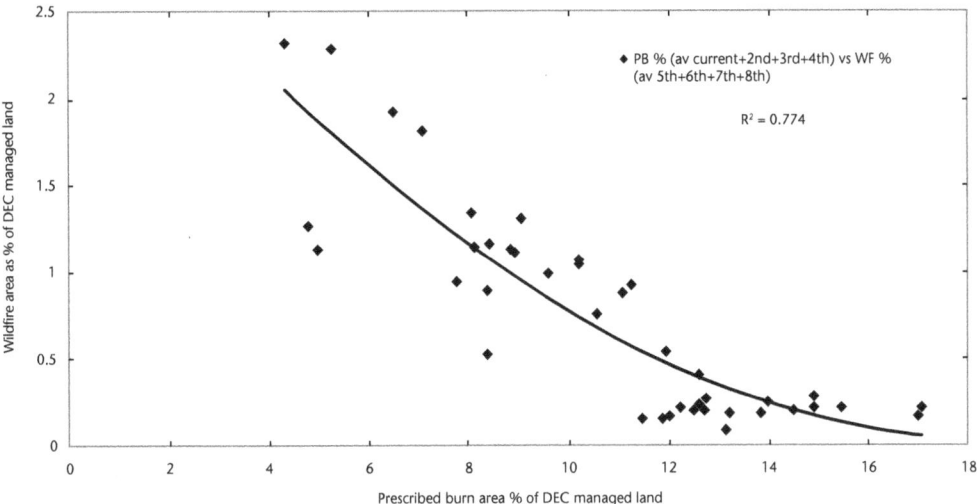

Figure 7.1.   Relationship between prescribed burn area (mean of 4 years) and bushfire area (mean of the succeeding 4 years) on land managed by Department of Environment and Conservation in the south-west of Western Australia (from Sneeuwjagt 2008).

burnt by bushfire for each succeeding 4 years, and then calculated these averages as percentages of the area of land managed by DEC. The correlation between bushfire area (percentage of land area) and fuel-reduction area (percentage of land area) is remarkable (Figure 7.1). The regression coefficient $R^2 = 0.77$ means that the regression between the two accounts for 77% of the variability in the data, and this is a remarkable result given the variability in fire weather, and the variability in fuel reduction. The regression shows that as average annual fuel reduction increases from 2% of the land to 16% of the land, the average area burnt annually by bushfires decreases from 2.1% of the land to a mere 0.1% of the land.

Sneeuwajgt's analysis is fully supported by subsequent work by Boer *et al.* (2009) in the forest of south-western Australia, again using a refined regression analysis (Figure 7.2).

The hypothesis that fuel-reduction burning decreases the intensity and rate of spread of subsequent bushfires is fully supported by these historical records and analyses for the forests of south-western Western Australia, and this is the best record we have for all of Australia. However, lessons from the west and clear statements from practitioners across southern Australia, have been only slowly, even grudgingly, accepted by some researchers and commentators. For example, submissions to the Council of Australian Governments Inquiry on Bushfire Mitigation and Management included the following:

'Cary and Bradstock (2003) argue that managers and experts can only speculate on critical levels of fuel-reduction burning required to meet particular management objectives in Australian systems because the relationship between the extent of fuel-reduction burning and the probability of unplanned high intensity fire is yet to be determined for any system.'

And also:

'For example, an extensive analysis of the contemporary fire regimes in the forests of south-western Australia (Gill and Moore, 1997), concluded that "since the introduction of prescribed burning to all areas of the Jarrah forest in the 1950s there appears to be little

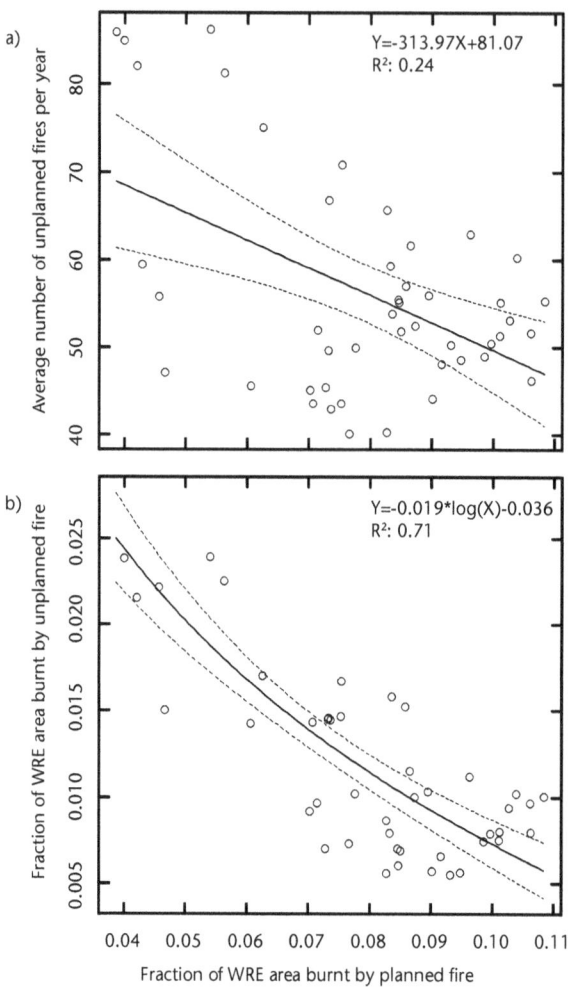

Figure 7.2.   Six-year running means of the annual number (a) and extent (b) of unplanned fires against 6-year running means of the annual extent of planned fire (1958–2003). Fire extent shown as fractions of the current surface area of the Warren Region Estate, south-west Western Australia (from Boer *et al.* 2009).

change in the unplanned-fire cycle based on total area of State forests and timber reserves but a widening of the interval if the basis was the protected area." The authors stated that they were unable to establish which basis was more appropriate.'

Sneeuwjagt's analysis (and much other evidence) clearly challenges these assertions. The evidence that fuel-reduction burning reduces the probability of unplanned, high-intensity fire can always be challenged at a point-by-point scale owing to variation in intensity of bushfires (and fuel-reduction fires) as they travel across landscapes and the many contributing factors to that intensity. However, it should be noted that the relationship demonstrated by Sneeuwjagt covers a wide range of levels of fuel-reduction burning and, as might be expected, represents a continuum of efficacy – fuel-reduction burning will be less effective when less of the landscape has been treated and more effective as the treated proportion increases.

There have been few fires larger than 10 000 hectares, no loss of lives and relatively little property damage in the south-west of Western Australia since 1961. Even the recent (December 2009) Toodyay fire (80 kilometres north-east of Perth) that destroyed upwards of 30 homes was relatively small (~3000 hectares). In contrast, more than 3 million hectares has burned by bushfires in Victoria in the period 2002–2009, with devastating loss of lives, property and live-stock. The Western Australian experience is often dismissed by the opponents of fuel reduc-tion as being irrelevant to the forests of eastern Australia. In response to this opposition, Sneeuwjagt comments:

> 'What differences exist between the south-east and the south-west that contribute to markedly different fire control records? In Western Australia the topography is relatively flat and undulating making it easier to undertake rapid attack on initiating fire. But there is no practical difference in the structure and flammability of the forest fuels, and the fire behaviour in long unburnt forests can be just as severe and destructive in the west (e.g. Dwellingup fires in 1961) as it is in the south-east forests.
>
> **The main difference between the south-west and south-eastern Australian States is the scale and frequency of fuel-reduction burning undertaken by the land management agencies.** In Western Australia between 6 to 8 percent of the forested crown lands are fuel reduced each year compared with less than 2 percent in other States.
>
> The level of fuel-reduction burning undertaken over more than 47 years in south-west WA have enabled fire managers to achieve a high level of protection to the community assets and natural values on and near the lands managed by Department of Environment and Conservation. There have been numerous examples where the fuel-reduction burning program has resulted in significant saves even under extreme weather conditions'. (Note: the emphasis is ours, and we have used 'fuel-reduction burning' in place of 'prescribed burning'.)

We do not have such thorough analyses of the effects of fuel-reduction burning for eastern Australia. The analysis by McCarthy and Tolhurst (2001) is of interest. At that time there were five Fuel Management Zones in Victoria's public forests, from Zone 1 (an 'asset protection' zone, an area managed to provide the highest level of strategic protection to human life, property and highly valued public assets and values) to Zone 5 (an exclusion zone in which there is no fuel-reduction burning).[22] The greatest effect of fuel-reduction burning in assist-ing with suppression of subsequent bushfires over the period 1990–91 to 1996–97 was in Zones 1 and 2 where fuel reduction had been most frequent (5 years in Zone 1 and 7 years in Zone 2), and this effect 'appears to be significant' according to the authors. In contrast there was little or no effect in Zone 3 (broad area fuel reduced mosaic – an area managed to provide an irregular mosaic of areas of fuel reduction) where the frequency of burning was about 11 years. The Victorian study is in general agreement with the far more detailed Western Aus-tralian study.

Another most critical consequence of fuel-reduction burning is in facilitating the success of 'first attack'. First attack or 'first strike' as soon as possible after a bushfire has started and been detected is the key factor in fire suppression – attack while the fire is small. The success of first attack in relation to the overall fuel hazard (surface, elevated and bark fuels) was assessed over the four summers 1991–1995 in Victoria (McCarthy and Tolhurst 1998; McCarthy *et al.* 2009). First attack by a small crew with basic equipment is 100% successful when the fuel hazard is low to moderate, even at a forest fire danger index of 100 (Figure 7.3). However, when the overall fuel hazard is very high to extreme (no fuel reduction over many years), a very

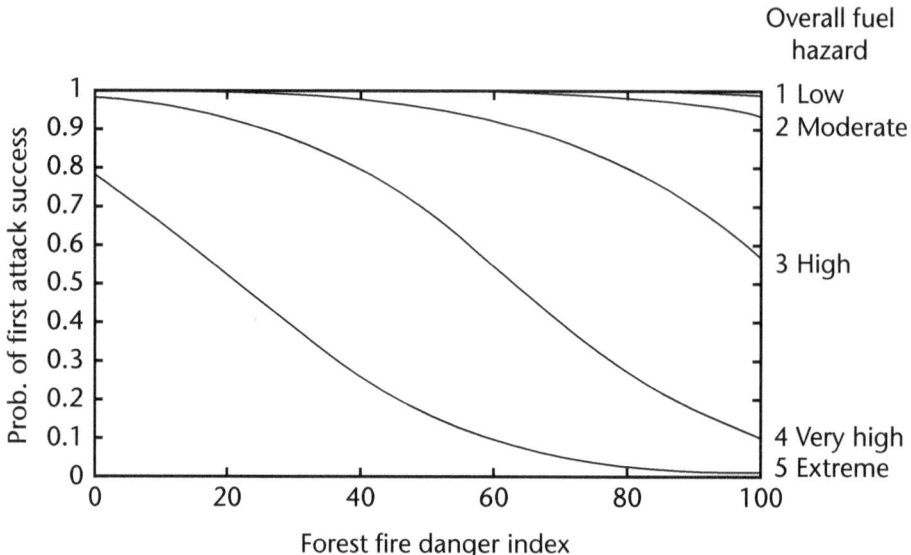

Figure 7.3.   The probability of success of first attack on a bushfire using a small crew (for example, six people) and basic equipment (for example, two 'slip-ons' – a water tank and a pump on the back of a four-wheel drive – and a bulldozer) as a function of both overall forest fire danger index and overall fuel hazard (from McCarthy and Tolhurst 1998; McCarthy *et al.* 2009).

much greater effort – a larger crew, more bulldozers and aircraft – is required for successful first attack.

Models

McCarthy and Tolhurst (2001) analysed 114 bushfires in Victoria in the 1990s in more detail. A simple model using 'time since last burn' (either by fuel reduction or by bushfire) versus probability of fuel reduction in assisting with suppression of bushfires was highly significant. Fuel reduction within the last 1–3 years was the most effective in subsequent bushfire suppression, this effect decreasing rapidly 7–8 years after fuel reduction to low probabilities 12–14 years after fuel reduction.

Another, most detailed study from Western Australia is Project Vesta (Gould *et al.* 2007) where a very large dataset collected from 104 experimental fires was exhaustively analysed. A model of fire spread was constructed using the strongest relationships from these analyses:

$$R = f(U, M, S, NS_s, NS_h)$$

where R = potential rate of spread.

U = mean wind speed at 10 metres height in the open.

M = moisture content of dead fine fuel.

S = hazard score for surface fuel. Surface fuel is litter on the forest floor, made up of leaves, twigs and bark of overstorey and understorey plants, in varying stages of decomposition. Hazard score varies from 0 = no surface fuel to 4 = litter depth of 25 centimetres.

NS_s = hazard score for near surface fuel. Near surface fuel includes grasses, low shrubs, creepers and collapsed understorey usually containing suspended leaves, twigs and bark from the overstorey. Hazard score varies from 0 = absent to 4.

NS_h = height of near surface fuel.

The outcome is a predictive model for fire spread in dry eucalypt forest. Furthermore, observations from the experimental fires confirm that spotting of firebrands ahead of the fire is 'intimately linked to the behaviour of the convection column' that forms under certain conditions. A second model has been developed from previous work to predict spotting distances in stringybark forests (for example, messmate, *Eucalyptus obliqua*, in eastern Australia and jarrah, *E. marginata*, in the west). As for the effectiveness of fuel-reduction burning, Gould *et al.* (2007) conclude that:

1. *'Hazard reduction . . . will reduce the rate of spread, flame height and intensity of a fire, as well as the number and distance of spot fires by changing the structure of the fuel bed and reducing the fuel load.*
2. *The persistence of this effect will be determined by the rate of change in fuel characteristics over time. Even when the surface fuel and understorey layers have stabilised, the hazard score rating of fibrous-barked trees will continue to increase and will increase the difficulty of suppression.*
3. *Stimulation of understorey shrub regeneration after burning will not increase the rate of spread of a fire until such time as a significant near-surface fuel load accumulates.*
4. *Younger fuel produces fewer firebrands because fire intensities are lower and less bark is consumed than in older fuel types. Reducing bark hazard from a hazard score of 3 to 2 by prescribed burning reduces the intensity of firebrands by threefold'.*

Finally, McCarthy and Tolhurst (2001) constructed a simple relationship aimed at predicting the probability that a previous fuel-reduction burn will slow down a subsequent bushfire. Probability (0–1) was calculated as a function of *overall* fuel hazard and Forest Fire Danger Index (FFDI, 0–100). Although the relationship was highly significant, we are not certain how the constants for the two parameters were calculated. However, with a low overall fuel hazard of 2 tonnes per hectare of surface fine fuels, the probability of slow down is above 0.8 for FFDI of 50 or less, decreasing to about 0.5 for FFDI = 100. As the overall fuel load increases to extreme (37 tonnes per hectare including 20 tonnes per hectare of surface fine fuels), the probability of slowing a bushfire is a mere 0.05 at FFDI = 25, reducing to zero at FFDI = 100.

Process-based computer simulation models have also been used to assess the probable effects of fuel reduction on subsequent bushfires, and here we use the results of a simulation for the buttongrass (*Gymnoschoenus sphaerocephalus*) moorlands of south-western Tasmania (King *et al.* 2006). In brief, three fuel-reduction strategies were modelled: (1) a proposal by fire managers to burn 3% annually of buttongrass moorland in a strategic way to protect fire intolerant vegetation; (2) a 'deterministic' planning of fuel reduction, with model blocks of the moorland burned in a defined sequence so that they differed by one year since burning; and (3) random burning of blocks so that some were burned more frequently than others.

The computer simulation of the mean annual reduction in area of bushfires with increase in fuel-reduction burning (Figure 7.4) for the buttongrass moorlands in Tasmania is remarkably similar to the actual decrease in area of bushfires with increase in fuel reduction in public forests and lands in south-western Australia (Figure 7.1). That is, there is a strong decrease in area of bushfires as annual fuel reduction increases from 0 to 20% of the land. The initial, steep decrease in bushfire area for this model requires about 5–6 hectares of fuel-reduction burning for each hectare of bushfire reduction, similar to that shown by Sneeuwjagt's (2008) analysis for south-western Western Australia. As might be expected, the decrease in bushfire area was greater under deterministic burning than under random burning, and the effect of

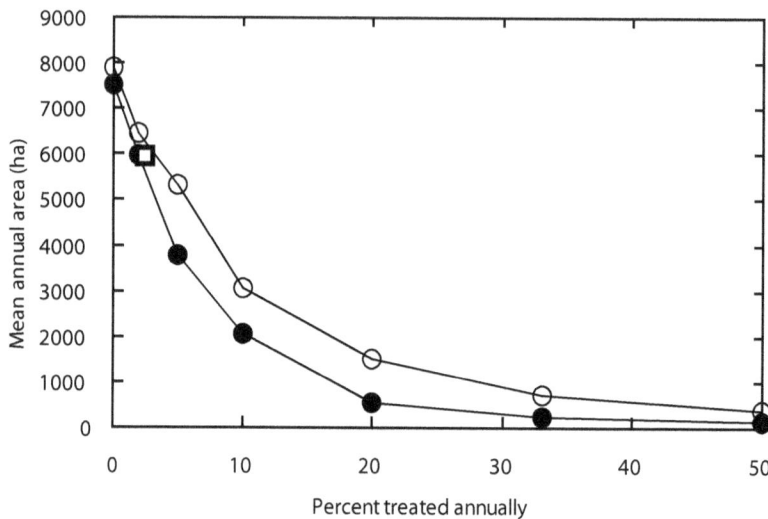

Figure 7.4.   Output of a simulation model showing the relationship between the mean annual area burnt by bushfires and the annual area of fuel-reduction burning in the buttongrass moorlands of south-west Tasmania (from King *et al.* 2006). The various, simulated treatments were: strategic burning of 3% annually (the single square); 'deterministic' burning of model blocks in a defined sequence so that they differed by one year since burning (closed circles); random burning of blocks so that some were burned more frequently than others (open circles).

strategic burning of only 3% annually was relative small and not significantly different from the other two treatments.

Remote sensing

The Joint National Park Service (NPS) – US Geological Survey (USGS) National Burn Severity Mapping Project was established to provide a long-term assessment, at the landscape level, of the magnitude of ecological change caused by fire.[23]

The methodology uses Landsat 7 satellite images[24] before and after a fire. From these images, a 'normalised burn ratio' (NBR) is calculated as the ratio of near-infrared spectral reflectance to mid-infrared spectral reflectance. The 'differenced normalised burn ratio' (*d*NBR) is then calculated:

$$d\text{NBR} = \text{NBR}_{\text{pre-fire}} - \text{NBR}_{\text{post-fire}}$$

Because of differences in spectral reflectance in the near-infrared and mid-infrared changes between green vegetation and burnt vegetation (especially blackened soils), the difference in their ratios before and after a fire (*d*NBR) is closely related to fire intensity. Higher values of *d*NBR are associated with higher fire severity, and this association has been thoroughly tested and verified by ground observation.

The effectiveness of fuel-reduction burning was assessed for two fires in ponderosa pine forests in Arizona using Landsat 7 images provided by NPS-USGS (Finney *et al.* 2005). The fires in 2002 totalled 186 000 hectares; they burned during record drought conditions and extreme fire weather, with wind gusts of 40 kilometres per hour, relative humidity less than 10% on most afternoons, and moisture content of fuels calculated at less than 4%.

Despite the extent of the fires, and despite the record drought and the extreme fire weather, Finney *et al.* (2005) found the following statistically significant results:

- Fire severity decreased in areas where fuel had been reduced up to 9 years before the fire.
- Fire severity increased with time since fuel reduction.
- Fire severity decreased with the number of repeated fuel reductions.
- Fire severity decreased with increasing size of the fuel reduced area.
- Fire circumvented fuel reduced areas.

These results confirm, at least for this vegetation type, that strategically planned fuel-reduction burns, of sufficient size and repetition, mitigate both the severity and size of bush-fires, and that this outcome is statistically significant.

Are we approaching a scientific consensus on fuel-reduction burning?

Fuel-reduction burning has, as we have shown, been a controversial topic not just among those with interests in the management of public land but among fire ecologists. During the 6 days of hearings on Land and Fuel Management (17–24 February 2010) at the 2009 Victorian Bush-fires Royal Commission, however, there was a surprisingly high level of agreement among scientists and among an 'experts forum'. Western Australian fire experts Lachie McCaw and Rick Sneeuwjagt gave evidence before the Commission, and the seven experts assembled for the forum covered a wide range of scientific advocacy on matters of fuel reduction.

The experts' forum reached the consensus view that between 5% and 10% of public land should be burned for fuel reduction each year. Given that the public forest estate is 7.7 million hectares, this gives an annual area to be burned of 385 000–770 000 hectares, or 3–6 times the 130 000 hectares that is currently burnt each year.

Following the week of evidence before the Commission on Land and Fuel Management, proposed recommendation coming from Counsel assisting the Royal Commission[25] included the following:

- *'It is essential that the DSE annual performance of planned burning of the public land estate be measured against a state-wide target of hectares burned.*
- *The annual target for planned burning should be between 5% and 10% of the available public land estate with an immediate goal of achieving a minimum annual rate of 385 000 hectares (5%) of planned burning.*
- *The total number of hectares annually treated by planned burning should be set out in the annual report of the Department of Sustainability and Environment'.*

Similar consensus among experts was reached in a case in the Supreme Court of Western Australia. The issue was whether the then Department of Conservation and Land Management had a duty of care to ensure that smoke from fuel-reduction burning does not cause damage to nearby winegrowers by tainting the grapes. The judgement[26] notes that an experts' conferral prior to trial agreed to the following propositions (among others):

'(a)   Prescribed burning is a necessary part of public land management to meet biodiversity, economic, silviculture, water and community protection objectives.

  (b)   Prescribed burning is the only effective way of reducing the potential impact of wildfires at a landscape level.'

It seems that more consensus is reached under the rigours of the legal system than is reached in the scientific literature at large.

Summary

The term 'fuel-reduction burning' is self explanatory: burning in a strategic and controlled way results in a reduction of forest fuel, depending on the prevailing conditions of the fuel and the weather. The amount of fuel is reduced so that the spread and intensity of bushfires is reduced, thereby greatly facilitating control and suppression.

The quantitative effects of fuel reduction on subsequent bushfires are increasingly documented as we move from simple observation to more refined models of fire spread and behaviour. McCaw *et al.* (2008) note that earlier models under-estimated the rate of spread of bushfires, thereby under-estimating the benefits of fuel-reduction burning. New models of fire spread for open eucalypt forest will enable more precise estimates of these benefits.

The Western Australian evidence discussed in this chapter (Sneeuwjagt 2008) is indisputable. Not only is the area of bushfire highly and negatively correlated with the area of previous prescribed burning, but the incidence of lightning-caused fires is less where the fuel is less than 5 years old (McCaw *et al.* 2008).

For fuel reduction to be effective in eucalypt forests, it must be repeated regularly. If we can control the fuels, then our job in fire suppression is greatly facilitated. All of the evidence presented here (for example, Figure 3.3) shows that once the surface fuel accumulates above 10 tonnes per hectare, a bushfire will be difficult to control even under moderate conditions.

A bushfire burning under extreme summer conditions creates its own weather and forms a significant convection column. Firebrands carried up in the column can be carried kilometres ahead of the main fire. All of the forest, except for the trunks of the trees and their larger branches, becomes fuel for the fire. Fuel-reduction burning will influence the spread of these fires, but it will not stop them.

Fuel-reduction burning must be acknowledged for what it is: a part of a program of fire management that includes fire prevention (fuel reduction, maintenance of roads, tracks and firebreaks), fire suppression (ground crews, tankers, bulldozers and aircraft), and a professional, aggressive and immediate first-strike capability.

Appendix: Contrasting approaches: Western Australia versus New South Wales and the Australian Capital Territory

We have used examples from Western Australia (WA) in many places in this text. We have also given some weight to examples from Victoria – another state that knows the risks posed by uncontrollable bushfires. We think it instructive to take a brief look at the approach to fuel-reduction fires in New South Wales (NSW) and the Australian Capital Territory (ACT).

The NSW approach is perhaps the most controversial of all states. Given that NSW surrounds the ACT, together with the small size of the ACT itself, fire-planning and approaches to fuel reduction are closely aligned. In terms of their flammability, fire-prone nature, and the risks they pose to life and property, large areas of forest in NSW and the ACT are not far behind Victorian forests – even those forests close to Melbourne that burnt so fiercely in 2009 (see Chapter 1). NSW and the ACT differ clearly from WA and other states in many aspects of the application and planning of fuel-reduction fires.

To begin, we might contrast simple statistics in area treated with fuel-reduction fires. Despite a larger forest estate in NSW/ACT than that in WA, much less is treated in NSW. In Chapter 2, we noted the almost total lack of use of fuel-reduction fires in the ACT. This speaks for itself and is not supported, as far as we are aware, by any 'unique' attributes of the vegetation or landscape.

The trend of state governments over the past 10–15 years to change tenure of forested land from state forest to national park (see also Attiwill and Adams 2008) has been accompanied by a significant decline in the use of fuel-reduction fires in NSW (see Jurskis *et al.* 2003). The area treated with fuel-reduction fires in NSW is scarcely more than that in Victoria (Figure 7.5), even though the area of forest in NSW suitable for fuel reduction is at least three-fold that in Victoria (>21 million hectares in NSW versus ~7 million hectares in Victoria); the comparison allows for the subtropical forests of the north coast of NSW where summer rain is significant, are so are difficult to burn.

In part, the NSW approach can be summed up by the rise to ascendancy of national parks over state forests. Where the state forest approach had been to assess direct evidence of ecological effects of fuel-reduction burning at an *operational scale* (as exemplified by the Eden Burning Trials), the Parks approach is to generate models based on studies of the effects of *fire regimes* (insofar as these can be assessed), largely concentrated in the Sydney basin. The current NSW/ACT approach to assessing fuel reduction is summed up by a recent *Sydney Morning Herald* report[27] of comments by the Rural Fire Service assistant commissioner, Rob Rogers:

> *'We no longer measure the amount we burn in hectares because it is not the best measure of hazard reduction. You can burn 100 hectares 100 kilometres from the nearest property or you can burn one hectare near to the property and get a better result in terms of safety.'*

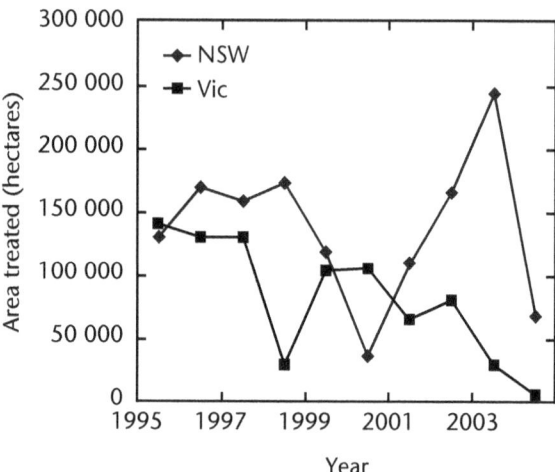

Figure 7.5. Area of forest treated with fuel-reduction fires in New South Wales and Victoria (Department of Environment, Water, Heritage and the Arts 2006).

The Canberra fires of 2003 and the Victorian Bushfires of 2009 exemplified problems with 'burning over the back fence' approaches to fuel-reduction treatments (see Box 9.3). Fires can develop sufficient momentum in terms of intensity and rate of spread and, especially if they become crown fires, that they can overwhelm local defences.

The approach of the NSW National Parks and Wildlife Service is explained in the 2004 guidelines for application of fuel-reduction fire – '*Guidelines for Ecologically Sustainable Fire Management*' (Kenny *et al.* 2004). This document is remarkable for several features.

First is its title. In contrast to globally accepted definitions of ecological sustainability (see Foreword and Chapters 1, 5, 6 for examples of other attributes of ecological sustainability), these guidelines consider only biodiversity. To make the point more clearly, consider the enormous Kosciusko National Park. This Park encompasses much of the Snowy Mountains – arguably the most important water and carbon reserve in Australia. Water from the Snowy Mountains is a significant proportion of that entering the Murray–Darling basin and, along the way, provides hydro-electric power. The soils of the high country are amongst the most carbon-rich of all Australian soils and clearly amongst the most susceptible to changes in temperature. Both water and carbon cycles are strongly influenced by fire (see Chapters 6 and 7) and are part of every definition of ecological sustainability. Yet these guidelines are silent on them.

Secondly, the NSW guidelines are clearly written with an assumption that the most important aspect is a need to avoid 'adverse' fire regimes of 'high frequency fire', albeit that 'high frequency' is never clearly defined. For example, on page 10: 'High frequency fire resulting in the disruption of life cycle processes in plants and animals and loss of vegetation structure and composition' has been listed by the NSW Scientific Committee as a key threatening process under the Threatened Species Conservation Act 1995. And on page 11: "Fire regimes may be adverse (beyond the limits of tolerance) for individual species at certain points in a community but not at others." Nowhere do Kenny *et al.* (2004) mention that what is adverse for some species can be beneficial for others. In contrast to the liberal use of the term 'adverse', the word 'beneficial' is used just once. Even the word 'favourable' is applied mostly to mean 'not adverse'. On page 30, Kenny *et al.* (2004) seek to quash any suggestion that heterogeneity of intensity within fires is beneficial to biodiversity (e.g. Chapter 8): '"Patchy" fires are not automatically beneficial to biodiversity.' Not only has this statement been directly challenged with robust data from fuel-reduction trials at Eden (e.g. Penman *et al.* 2007), it makes little logical sense. In fact, patchy fires are inevitable under controlled conditions; low-intensity fuel-reduction burns are inevitably patchy (Burrows 2008).

Thirdly, and perhaps more than anywhere else in Australia, the guidelines are based on theories of succession and the assumption that fire is a 'disturbance'. The guidelines set a 'lower threshold'[28] as: 'The minimum interval (based on the minimum maturity requirements of plant species sensitive to extinction under frequent fire regimes) is the shortest inter-fire interval needed to avoid **any localised declines or losses of species** as a result of too frequent fire' (Kenny *et al.* 2006). The emphasis is ours.

Given we know that fire changes species composition, with some species reappearing quickly after fire but giving way to others over time, this statement is open to considerable interpretation. Over what time scale will any decline (let alone loss) of species be measured? 1 year? 10 years? 100 years? Over what spatial scale? The area of a single fuel-reduction fire? A whole valley? Elsewhere in Chapter 7 we comment further on the use of single species, 'shortest inter-fire intervals' as the basis for setting inter-fire intervals for communities.

Some of the authors of studies used as the basis of the guidelines and the overall approach of NSW Parks acknowledge the tenuous nature of their interpretation of the data. For example, Cary and Morrison (1995) and Morrison *et al.* (1995) spend some time discussing the considerable numbers of caveats that need be considered in interpreting their data, and most of the supposed 'evidence base' of risks to plant diversity is highly contestable. We have commented elsewhere (e.g. Attiwill and Adams 2008) on the 'reference' problem (again, see also Chapter 7) for studies of the effects of fire (or fire regime) and its corollary that choosing an inappropriate reference automatically privileges some species while disadvantaging others. Jurskis *et al.* (2003) questioned much of the empirical data that was later cited by Kenny *et al.* (2004) as being conclusive evidence of the risks to biodiversity throughout the state. Even the Nature Conservation Council of NSW has commented: 'It should be noted that when thresholds are determined for a very broad geographic area, as is the case in the Kenny *et al.* (2004) guidelines, lower thresholds may be set by species **which are not found in particular local areas**, and thus may be higher than if local species only were considered' (Watson 2006).

Again, the emphasis is ours and in Chapter 8 we highlight the work by Vivian *et al.* (2010) as providing sobering notes of caution in applying such approaches.

The process for approval and implementation of any given fuel-reduction fire (see Table 7.1) also deserves comment. Although each step may be reasonable in isolation, when combined they comprise a task of Everest proportions for any person responsible for implementing fuel-reduction fires. For example, the (District) Bushfire Management Committees and their consultative processes all sound good in theory – bringing together the parties responsible for an area affected by fuel-reduction fires. There are obvious benefits including making sure agencies such as the police and ambulance service and local government are informed and can bring to the table relevant issues and concerns. These processes also provide a hearing for objections from individuals and local groups.

However, an extraordinary part of the process is that the Nature Conservation Council has been granted a formal say, enshrined in legislation. From their website (Nature Conservation Council 2010):

'Since the passing of the Rural Fires Act 1997 and the Fire Brigades Act 1989, the NCC has had responsibilities under these pieces of legislation to appoint conservation representatives to the Bushfire Coordinating Committee (BFCC) and Rural Fire Service Advisory

Table 7.1. Process for fuel-reduction burning in New South Wales (from New South Wales Department of Environment, Climate Change and Water 2010).

Steps	Tasks
Update databases	Update fire history on GIS generate year Record results of previous years programs
Review burn prescriptions	Review prescription burn requirements on GIS and set priorities Undertake GIS threat analysis Undertake GIS fire regime evaluation
Fuel sampling	Undertake fuel sampling in areas for proposed burns and collate data
Prepare burn proposals for DBFMC's	Submit prescribed burning proposals to District Bush Fire Management Committees Seek approval by DBFMC's
Review of environmental factors	Prepare Hazard Reduction Certificate or review of environmental factors (REF) Undertake field assessments if required Seek approval for HR certificate or REF
Risk assessment	Undertake field inspection to assess risks with implementing burns
Prepare operations plans	Prepare operations plan Seek operations plan approval
Implement burns	Identify appropriate conditions to implement burns by monitoring weather conditions Notify RFS fire control centres, emergency service agencies and neighbours Organise resources including RFS crews
Post burn assessment	Undertake post burn assessments for effectiveness
Monitoring and reporting	Report to DBFMC's on the success of the prescribed burning program. Maintain internal databases and reporting protocols Percentage of the total area actually treated

Committee (RFSAC) as well as to the Bushfire Management Committees (BFMCs) around NSW.'

The Institute for Public Affairs is understandably less than impressed:

'The NCC has been given a privileged position in the formulation of policy for mitigation of fire in NSW. It has sat on the central policy-making committee. It has a seat on the central and local co-ordinating committees. Yet the NCC represents a negligible constituency. The bulk of its funding is provided by taxpayers. Despite this, it has a position on a par with the politicians and government departments responsible for policy and with the State Forests and Parks agencies that manage the vast areas of public land. It has influence beyond the tens of thousands of private landowners and managers who are responsible for managing most land in the State, who have enormous collective experience and depend on the land for a living. The NCC suffers no consequences from the failure of the policies it advocates. It does not have to clear up in the aftermath. It is profoundly unfair and antidemocratic that

a single unrepresentative group should have influence of this kind over the interests of those with most at stake.' (Hogget and Hogget 2004).

The *Sydney Morning Herald* report cited above also quoted Assistant Commissioner Rogers of the Rural Fire Service as saying: 'We have a good work relationship with the environment movement'.

The official position of the Rural Fire Service (RFS) is not surprising given the circumstances of how fuel-reduction fires are planned and implemented in NSW, the tremendous growth in area of land managed as national parks, and the Kenny *et al.* (2004) guidelines. Indeed, the RFS would struggle to do any fuel-reduction burning unless it had a 'good relationship' with the NCC. Likewise, the division of land to be treated into smaller and smaller parcels consumes more and more resources per hectare and occupies staff for the small windows of opportunity dictated by the weather and by safety and other prescriptions. It is hard to see how the staff responsible for implementing fuel-reduction fires in rural NSW can have the flexibility, time and resources needed to make both good decisions as to when to start any given fire, and to determine how large should be the areas burnt.

A last word might belong to Noel Pearson of the Cape York Institute and well-known advocate of the rights of rural and Indigenous Australians. In commenting[29] on recent Queensland legislation that he and many others view as a major concession to city based environmental groups, to the detriment of those that live in Cape York, he wrote:

'The charade of participatory democracy can be seen in every region of the State where there are networks of "catchment management groups" and "natural resource management groups". Farmers, local communities, indigenous representatives and shire councils sit down with State Government bureaucrats and representatives of green groups and supposedly work out consensus solutions to land use and environmental management. But what the mug stakeholders from these communities do not realise is these processes are tokenism. The real decisions are made in Brisbane. The people who actually live in these regions and who strive to make a livelihood out of the land, are reduced to being bit-part "stakeholders", while the real players are those cutting the deals in Brisbane."

We do not suggest any 'deal' was ever made in relation to the NSW approach to fuel reduction. On the other hand, we will leave it to the reader to question the decision of the government (and senior bureaucrats of the day insofar as they provided advice to government) to allow one interest group, and essentially a city-based interest group, a privileged position in something as fundamental to the lives and livelihoods of so many, as fuel reduction fires. The NSW/ACT situation seems to us to epitomise our comments elsewhere that fuel-reduction fires will continue to be a political issue so long as Australia remains an intensely urban society and as the divide between city and country grows.

8

Fighting fire with fire: II. Fuel-reduction burning and diversity

Arguments about the effects of fuel-reduction burning on biological diversity rage, both among ecologists and within the wider community. Large amounts of work up to 2003 were reviewed and discussed at length in books by Bradstock *et al.* (2002), Abbott and Burrows (2003), Bowman (2000) and Mackey *et al.* (2002) and in reports (e.g. Gill 2008). These books and reports draw on the many studies of the effects of fire on individual species and plant communities. There have been only modest additions of new, published research since 2003 on the effects of fire on individual species – that is, our knowledge at the species level advances slowly, as it must.

More extraordinary is the almost total lack of any published research on the effects on diversity of the major, extensive and often high-intensity bushfires that have swept though much of south-eastern Australia since about that time. We say extraordinary, because it seems self-evident that 'all fires – planned or unplanned – should be treated as opportunities from which the public, scientists and managers can learn about fire impacts' (Burrows and Abbott 2003). While, in 2008, the *International Journal of Wildland Fire* published a series of papers dealing with large fires, contributions relevant to southern forests in Australia largely iterated previous discussions of fires and diversity (e.g. Bradstock 2008; Gill and Allen 2008) rather than presenting data of the effects of the 2003 or 2006–07 fires on biodiversity. To our knowledge, there has been no large-scale analysis of the number of animals killed by those bushfires or analysis of the effects of the fires on plant diversity or community composition. This is not a recent trend.

Major bushfires in south-eastern Australia since 2002 include:

- 181 400 hectares, Big Desert Wilderness Park and Wyperfeld National Park, Victoria, 2002 (almost entirely public land)
- 1.5 million hectares, New South Wales, 2002–03 (including two-thirds of Kosciuszko National Park, and a considerable area of the southern Alps)
- 1.3 million hectares, Alpine Fires, Victoria, 2002-–03 (almost entirely public land)
- 160 000 hectares, Canberra fires, Australian Capital Territory (ACT), 2003 (almost 70% of ACT, including 126 000 hectares of plantation forests and nature reserves, the rest rural land)
- 145 000 hectares, Eyre Peninsula, South Australia, 2005
- 160 000 hectares, Grampians National Park (50% of the entire Park burnt) and Anakie, Victoria, 2005–06 (60% on public land)
- 1.2 million hectares, Great Divide fires, Victoria, 2006–07 (1 million hectares of public land
- 430 000 hectares, Black Saturday fires, Victoria, 2009 (including 70 national parks and other reserves).

In the seven years 2002–2009 in Victoria alone, some 3 million hectares of public land – almost 40% of the state's public land – was burnt by bushfire. This amounts to about 430 000 hectares of our land burnt at relatively high intensities each year. We can only guess at the number of animals killed by these fires, with little opportunity to find refuge. Bushfires over such extensive areas, at intensities so high that over more than 50% of the area the tree crowns were completely removed, create uniformity over excessively large areas, not diversity.

In contrast, over the 8 years 2000–2008, the target area for cool, fuel-reduction burning in Victoria averaged 121 250 hectares per year, but the actual area burnt was only 95 262 hectares per year, or less than 1.5% of the area of public land.

The point we are making here is that we have had a great deal of research on the effects of fuel-reduction burning on diversity but, by comparison, little research on the effects of bush-fires (as is clear from Bradstock's discussion in 2008). We hasten to add that we do not fit within Bradstock's anecdotal perception as people who suggest that 'large fires are abnormal or unnatural': that is not our argument – rather, we argue that in this day and age of increasing ecological knowledge and advanced technology, large bushfires over extensive landscapes are *disasters* that make no ecological or social sense.

The nature of research on fire regimes and diversity in southern forests

We have already made the point that, surprisingly, there has been very little new information, or even knowledge, of the effects of fire regimes – bushfires and fuel-reduction fires – on bio-diversity in southern forests in the last 6–7 years. For example, Gill's (2008) extensive report lists more than 370 references, but not one single article reporting new (post-2003) empirical evidence as to the effect of prescribed fire on individual plant or animal species in forests. Instead, models that predict outcomes of this or that fire regime for biodiversity have become the focus of many fire researchers. Of great concern is that few, if any, of the predictions of such models are ever tested in the field. This is the case even when the authors make dire pre-dictions about losses of diversity if this or that fire regime is adopted or, worse, if this or that regime is adopted universally. Given the evidence of the diversity (scale, timing, intensity and patchiness) of fuel-reduction burning in the last 40–50 years, the lasting benefit of many, if not most, of the published modelling studies would seem very slight indeed.

On the other hand, we do have some added information about the effects of fire regimes on insects (e.g. York and Tarnawski 2004) and fungi (e.g. Osborn *et al.* 2003), and some of these southern forest studies have been complemented by studies of fungi in very frequently burnt forests in Queensland (e.g. Bastias *et al.* 2006a, 2006b; Anderson *et al.* 2007). We have refined our knowledge of the effects of fire regimes on some aspects of species assemblages (e.g. Pekin *et al.* 2009). We know more about the effects of fire regimes on the composition of grassy understoreys in woodlands (e.g. Prober *et al.* 2007; Watson *et al.* 2009), but not in forests. We know more about the effects of the *absence* of fires (e.g. Penman *et al.* 2009; Close *et al.* 2009).

Life history attributes

Overwhelmingly, the planning of fuel-reduction fires in relation to biodiversity is still reliant on generalisations based almost entirely on life history attributes. Among the many attributes of plants that respond to fire and that contribute to the survival of a species in the long term, life history attributes have had the most attention owing to their ease of observation. Species are thus usually listed according to their 'known' attributes in relation to the mode of reproduction (sprouting or from seed) and the minimum time required between fires for plants to be able to reproduce (often referred to as the juvenile period) and survive. The classic example is that provided by Gill and Bradstock (1992) in their 'National register for the fire responses of plant species'. The great majority of individual species assessments on this register are not backed by experimental evidence (e.g. experimentally tested at multiple sites) – nor could they be given the large numbers of species involved. There has been insufficient time and resources for experimentation to establish these attributes exactly, despite many calls for resources to be made available. They are instead collections of field observations by many different observers, with a few confirmatory experimental studies. This ensures that many, even most, species-level assignments of minimum fire-return intervals are conservative. That is, most of the 'minimum' fire-return intervals for the proposed individual species have not been rigorously tested, and the innate tendency of scientists to err on the side of caution ensures that without specific evidence to the contrary, they will 'round up' and suggest greater minimum fire-return intervals rather than 'rounding down' and shorter intervals.

Clearly, use of lists of life history attributes (especially when used in models) ought to be careful and considered in light of the uncertainty about any given species or even sets of species. Unfortunately, once a species is assigned a minimum fire-return interval (noting the above trend for such intervals to be 'best guesses'), then that value is widely regarded as fact in fire planning and policy. Moreover, application of such information (for example, adoption for a given *community* of a minimum fire-return interval that matches that of the *species* with the longest fire-return interval – often referred to as a 'threshold') can result in policy that clearly privileges less common species that benefit from longer fire-return intervals, over more abundant species with shorter fire-return intervals.

Dr Kevin Tolhurst has been instrumental in establishing life history attributes as part of the planning for fuel-reduction fires in forests in Victoria (e.g. McCarthy and Tolhurst 1998; McCarthy *et al.* 2003). The 2003 report contains a worked example of the application of the approach for the Grampians National Park and also includes caveats that need be considered alongside life histories (e.g. consideration of other attributes such as soil erosion), in setting fire regimes.

As a partial solution to the problem of imperfect knowledge, Gill and McCarthy (1998) argued for increased emphasis to be given to the 'indicator value' of 'serotinous seeders' – a class of plant that depends for its survival on seed stored in the canopy and only released after fire. Serotinous species typically need long fire-return intervals, often the longest for any given plant community. Gill and McCarthy wrote: 'the 'indicator' value is that if these species survive, it is assumed that other, more resilient, species will survive'

It is arguable that an approach similar to that described by Gill and McCarthy has been applied in parts of south-eastern Australia (e.g. montane and sub-alpine areas) and that the result has been long fire-return intervals with resulting dominance by woody shrubs, including serotinous indicator species. This is seemingly at odds with the majority of descriptions of that vegetation at the time of European settlement and, given clear increases in fuel loads and flammability, at odds with bushfire mitigation responsibilities of land managers. When coupled with the lack of certainty and in-built conservatism associated with assigned minimum fire-return intervals, it is not surprising that questions are raised as to the validity of the approach, as applied in practice at the landscape scale. A reasonable suspicion is that the compounding effects of 'rounding up' and uncertainty around minimum fire-return intervals for individual species, and adoption of highly conservative minimum fire-return intervals for whole communities (and even fire-exclusion policies for some areas), have resulted in unnecessarily large fuel loads with no increase in the security of plant diversity.

These things said, the underpinning and updating of vital attribute lists or databases, not by further speculation or modelling, but by proper experimentation and measurement, should be a standing research priority for Australia. Only through proper evidence of such life history attributes will Australia have the knowledge from which wisdom can be developed.

Perhaps the outstanding example of the importance of the latter is the recent work by Vivian *et al.* (2009). Using as a reference, the New South Wales Flora Fire Response database (NFFRD; NSW National Parks and Wildlife Service 2002), Vivian and co-workers found extraordinary divergence with results from a field assessment of how plant species recovered after the 2003 fires in the Brindabella Ranges. For example, the database contains conflicting information about species responses to fire. Vivian *et al.* (2009) thus devised 'rules' for classi-fication of species when different observers have not agreed (e.g. as to whether a species is a seeder or a sprouter). 'Rule 3' (see Vivian *et al.* 2009), when applied, required that species would be classed as obligate seeders – the most sensitive to short inter-fire intervals. Among their many findings of disparity between the database and the field, they found that: 'Rule 3 overestimated the number of non-sprouters by over 500% and the number of post-fire seeders by over 150%.' Vivian *et al.* (2009) made a number of important points including:

- 'Overall, the database did not concur very well with the observed fire responses'.
- 'The results of this study showed that using species classifications from the NFFRD underestimated the dominance of resprouting by between 30% and 50% compared with using the field data.'

Although the life history approach (including the 'indicator species' approach) has intui-tive appeal, mainly for its simplicity, it can also be argued that it gives little or no weight to:

- other ecological and physiological plant attributes that are affected by fire and that have direct relevance to the long-term species composition of plant communities (e.g. rates of growth, herbivory, tolerance of edaphic conditions, etc.)
- other community attributes that are critical elements of ecology and are key parts of the ecosystem services provided by plant communities.

The issue of conservation management in a world of limited resources has been thoroughly analysed by Professor Hugh Possingham and colleagues (e.g. McDonald-Madden *et al.* 2008a, 2008b; Bottrill *et al.* 2009). These authors have argued strongly that a form of 'triage' is essen-tial in conservation planning. Attempting to manage for single species or genotypes is seldom successful and wastes valuable, and necessarily limited, resources that are better spent ensuring conservation of the broader population of a given species or groups of species. Kevin Gaston (2010) recently summarised much of the global thinking amongst conservation biologists on the issue of focus on 'rare versus common' species. He noted the importance of common species for provision of ecosystem services (soil formation, primary production and nutrient cycling) as well as, obviously, the importance of rare species.

Given the aforementioned problems with species attributes as recorded in databases, the case for basing minimum fire-return intervals (thresholds) on less common or even locally rare species, is weakened further by the often opposite and sometimes beneficial effects of fires on more common species and on other attributes (e.g. see Chapters 5 and 6) of the plants and communities in question.

What is the evidence that fuel-reduction burning results in loss of diversity?

In contrast to the results of modelling studies built on life history attributes, which often focus on risks posed by too frequent fire, there is no evidence of any loss of diversity when fuel-reduction burning is actually tested in the field using realistic fire-return intervals (for example, 5–10 years). Instead, the concluding paragraphs of papers reporting the effects of such empirical trials often include clear and straightforward commentary and suggestions for land management. For example, a recent report from the forests of the Eden Burning Study Area (EBSA, Penman *et al.* 2008):

'More commonly, prescriptions have been adopted to exclude fire from the habitat of certain threatened species, usually without data on the impacts of fire or fire exclusion on these species. High frequency fire regimes benefit early successional species, whereas adopting a fire exclusion policy is likely to favour late successional species at the expense of early successional species. Both approaches have the net effect of reducing biodiversity and a balance needs to be reached'.

Results and conclusions from such scientific, empirical studies are not always reflected in opinion pieces. Consider the broad generalisations in the following piece from Campbell (2008):

'Calls for more frequent and larger scale fuel reduction burning is a predictable response to worsening bushfire risk in drought and a drying, warming climate. However this also has major (natural resource management) implications. As mentioned earlier, in addition to affecting water quality and quantity, increasing fire frequency changes species composition and floristic structure, favouring species that like fire including weeds – many of which are highly flammable, thus increasing fire risk and hazards over the long term. Further, a warming, drying climate makes it much more difficult to undertake so-called 'cool' burns safely. The windows of optimum weather will be shorter and rarer, and more likely to be followed by periods of higher fire danger, thus leading to even more wildfires originating from prior fuel reduction burns'.

Another example: The Victorian National Parks Association (2009) uses a passage from the 1939 Stretton Royal Commission Inquiry into the 1939 Bushfires to illustrate a seeming capacity for fuel-reduction burns to produce perverse outcomes:

'They [the settlers] burned the floor to promote the growth of grass and to clear it of scrub which had grown where, for whatever reason, the balance of nature had broken down. The fire stimulated grass growth; but it encouraged scrub growth far more. Thus was begun the cycle of destruction which cannot be arrested in our day. And so today in places where our forefathers rode, driving their herds and flocks before them, the wombat and the wallaby are hard put to it to find passage through the bush.'

The Victorian National Parks Association follows this with the statement, in its Preliminary Submission to the 2009 Victorian Bushfires Royal Commission:

'We are not making a suggestion here that burning always produces more fuel' *(the emphasis is ours, to show our incredulity!)*

In 1939, Judge Stretton did not have a lot of scientific evidence to rely on; however, in the 70 years since his report, no one has shown, or even suggested, a decline in the numbers of wombats and wallabies. In marked contrast to Judge Stretton's (1939) comments, and with the benefit of scientific rather than anecdotal evidence, Jurskis (2005a) presents the case that natural fire regimes (lightning and Aboriginal burning) over tens of thousands of years 'stabi-lised eucalypt ecosystems so that they were self-sustaining', and he reviews the mounting, published evidence that the reduction in the frequency of low-intensity fires has resulted in, among other things, invasion by shrubby understoreys and an increase in extensive fires of high intensity. Jurskis writes:

> 'Changes in fire regimes have upset the ecological balance in many eucalypt ecosystems, and caused the health and predominance of eucalypts to decline . . . (We should) recognise the imposition of unnatural fire regimes as a disturbance . . . Many ecological imbalances could be remedied at a landscape scale by reintroduction of more natural fire regimes. Unfortunately (those) ecologists and environmentalists (who adhere) to philosophies of non-intervention or passive management have supported misconceptions and confusion based on traditional concepts of disturbance and succession . . . They have opposed any reinvigoration of (fuel reduction burning) in the landscape. This, together with controversies about the recent (2002–2003) widespread and disastrous fires resulting from non-intervention will make it difficult to implement practical solutions to decline of eucalypt forests'.

In summary, if there were evidence of changes in diversity or of extinctions of species due to fuel-reduction burning at intervals of 5 years or more, we would be outspoken in our demands for modification or cessation of fuel-reduction burning, depending on the evidence. But there is no empirical evidence, as far as we are aware. Rather, the counter-position to our argument comes from modelling based on limited empirical data, and from anecdotes and opinion. For example, some scientists argue there is evidence that fuel-reduction burning makes forests drier and thus more flammable. When critically assessed, that evidence is very weak – at best, there is data that the authors used to argue species from drier sites were tempo-rarily appearing on wetter sites that have been recently burnt – and suggestions that reducing leaf area *reduces* water availability in southern eucalypt forests fly in the face of the laws of physics and chemistry and decades of hydrological research. We do not know of a species that is now extinct because of planned and strategic fuel-reduction burning.

The study of the effects of fuel-reduction burning in the Wombat Forest (Department of Sustainability and Environment 2003) is important because it is among the most detailed and long-term studies in Australia. Furthermore, it was based in Victoria's extensive mixed eucalypt foothill forests where most of the fuel-reduction burning is done and where there is a significant rural–urban interface. We quote some initial studies on flora (Tolhurst 1996):

> 'No plant species was gained or lost from any treatment. This was also the finding of other studies in similar vegetation. No sclerophyllous plant has ever been reported as having been made extinct as a direct result of burning, but some species have been eliminated from local areas as a result of frequent fires (e.g. mountain ash and alpine ash were eliminated in some areas of the Central Highlands of Victoria when burnt in 1926 and again in 1939). The relative abundance of different species varied with time since burning depending on the rate of development and the method of persistence.'

The longer term results support our conclusion that fuel-reduction burning at intervals of 5 years or more has little to no effect on diversity, and we have included two major, unedited extracts from the final report of the Department of Sustainability and Environment (2003) as an appendix to this chapter. Of course, we recognise that this study considered only a tiny fraction of total diversity; the estimated numbers of species for Victoria (data from Victorian

National Parks Association 2009), for example, are 670 species of vertebrates, 2500 species of non-vascular plants, 4300 species of vascular plants, 15 000–30 000 species of fungi, and 50 000–80 000 species of invertebrates! And so we hear the Victorian National Parks Association: 'We support fuel-reduction burning, but we need more research because we have more than 100 000 species to look after!' No series of studies, no matter how multidisciplinary and well-staffed, will ever encompass this total diversity; indeed, it would be futile and a waste of resources even to contemplate it.

We take a position developed by Taylor (1990). Taylor's concept of a 'natural' forest is one which is recognised as supporting native vegetation by simple field observation and the application of conventional phytosociological criteria. In dealing with the term 'natural forest' in Australia, Taylor states that:

> 'The present equilibrium vegetation (in Australia) has not been 'isolated in time' from the pre-Aboriginal native vegetation of the late Pleistocene. It has descended from this late Pleistocene native vegetation through an unbroken sequence of autogenic (natural changes within the community) and allogenic (changes in the environment) successional responses to human-generated disturbance and other natural agents of landscape change.'

Taylor's concept rightly recognises the role of humans over tens of thousands of years and of natural agents of landscape change over millions of years in forming the natural forest. Regular fire, caused by lightning and humans, has been a major part of this unbroken sequence of change. We can readily identify the natural forest, and if humans have degraded the natural forest then we can take the steps to fix it. The concept of the natural forest is scientific; in contrast, popular terms such as 'pristine' are totally vague. What do people mean when they talk or write of pristine forests? Forests that have not been touched by humans? Or not been touched by humans in the 220 years since British invasion? Or not been touched by fire?

Planning fuel-reduction burning for diversity outcomes

Gill's (2008) report on fire and adaptive management for the Department of Sustainability and Environment, Victoria, follows the many current models that divide the landscape into increasingly smaller pieces. The logic is that a 'mosaic' of fire-regimes and fire histories is required to optimise diversity, and that this will be best achieved by fine-scale, planning for fire. There are real practical difficulties to implementing this approach. First, it requires an ever-expanding bureaucracy to cope with the increasing planning needs created by subdivisions of landscapes into thousands, even tens of thousands of small blocks. Even if divided into blocks of approximately 500 hectares, the design of the more than 14 000 fire prescriptions in Victoria and more than 40 000 in New South Wales, based on analysis of plant diversity at the block scale, would be a massive task, and probably well beyond the resources that will ever be available. And 500-hectare blocks are many fold larger than preferred sizes suggested by some proponents of this approach. The approach also has the corollary of needing large numbers of staff in a central office and withdrawal of resources from the bush. Given that there are a finite number of days in any given year suitable for fuel-reduction fires, any approach that ties up resources treating smaller and smaller areas of land will, in effect, be a form of limiting the total area that can be treated in any given time frame. Finally, the 'chequer-board' approach gives little value to the mosaic created by fire, in favour of a human-proclaimed, lines-on-a-map mosaic.

In the absence of new knowledge, and in light of the large amount of information and knowledge that points to the fire-adapted nature of southern eucalypt forests, one of the clearest policy needs for fuel-reduction burning is a thorough review of all areas declared 'fire-exempt' on biodiversity grounds. A second, scarcely less urgent, need is to review guidelines for the area of land to be burnt in any single prescribed fire, and the timing of that fire in relation to annual seasons (e.g. autumn, spring) that are often at least partly different now to what they were 20 years ago. The latter may require an overhaul of management structures, especially the allocation of responsibility and authority for prescribed fires.

An alternative to geometric block, model-based approaches is to return responsibility to locally based, forest and fire officers, and provide the policy framework that allows much greater flexibility in the size and timing of prescribed fires, right up to the ability to use prescribed fire at the landscape scale (e.g. at the scale of thousands of hectares). In contrast to theoretical models, there is no evidence of homogenisation, or for that matter, elimination of biodiversity, when fuel-reduction burning is used at this scale in the southern forest estate.

Finally, a form of triage could, even should, be applied to planning for fuel-reduction burning. Fire regimes should not be chosen or applied solely on the basis of ideal fire-return interval of any single constituent species in any community or even landscape, even if that species has some 'indicator value'. There are far too many uncertainties in any such value for it to be a practical tool and the risks to plant diversity, as well as the created bushfire risks, are too great. A better approach would seem to be to focus on ensuring diversity of fire regimes at the landscape scale, including the heterogeneity of fire-return intervals encompassed within fuel-reduction fires.

One of the great problems in planning of fuel-reduction burning is that too often, and inexplicably, government departments develop prescriptions that exclude fire from some habitats without evidence (Penman et al. 2008). The most common excuse, even justification, for such disregard for 'evidence-based decision making' is the wrongly applied 'precautionary principle'. In other words, if in doubt don't burn. That is not application of knowledge and it is certainly not wisdom. The illogicality of this approach is all too frequently made clear by bushfires that not only burn the fire-excluded habitat (for example, the large swathes of montane and sub-alpine forests in north-eastern Victoria and southern New South Wales), but do so at greater intensity than any prescribed fire has, or will at any time in the future, and as often out-of-season as in-season (August fires in coastal forests in New South Wales are an example of the former).

Variability of fuel-reduction burns

Penman et al. (2007) quantified the frequently made observation that even when a prescribed fire is set, not all of the area it encompasses actually gets burnt. A large proportion of randomly placed study plots in the Eden Burning Study Area escaped fire one or more times in a four-burn cycle. 'Prescribed burns were extremely patchy at both the coupe and the plot scale for all treatments' (Penman et al 2007). This result is supported by every major trial of fuel-reduction burning in Western Australia, Victoria and New South Wales.

The patchiness of fuel-reduction burns is critical for planners and users of prescribed fire. It is a vital counter argument to those who either have weak knowledge of what actually happens in fires or set out to try to influence policy to suit their particular view. For example, the submission of the Australian Centre for Biodiversity, Monash University to the 2009 Victorian Bushfires Royal Commission states:

'Moreover, natural ecosystems exist as mosaics and the capacity of plants and animals to persist depended on large, continuous swathes of habitat. European settlement has

fragmented these swathes so our native biodiversity is unduly constrained into small, isolated parcels of native vegetation within which mosaics are difficult to maintain.'

We find it impossible to reconcile this statement with the forest estate of south-eastern Australia that is neither fragmented nor small. The submission states further:

'Applying a single fuel reduction burning policy to Victoria's public lands will disadvantage a large proportion of the native biodiversity and reduce local and regional habitat diversity. Shifting toward more homogeneous landscapes through increased fuel reduction burning will be detrimental to the long-term conservation of biodiversity in Victoria.'

Again, we find it impossible to reconcile this statement with the fact that less than 2% of Victoria's public land has been burnt annually by fuel-reduction burning, whereas bushfires of much higher intensity than a fuel-reduction fire have burnt 40% of the land in the 7 years up to 2009!

Fuel-reduction burning does not homogenise landscapes. Fuel-reduction burning in the cooler months of the year inevitably leaves some areas unburnt and burns the rest at variable intensities. There is no evidence that the use of prescribed fire in southern forests has created uniformity and, moreover, there has never been a 'single fuel-reduction burning policy', nor has one ever been proposed. Commenting on the Eden Burning Study Area, Penman *et al.* (2007) writes: 'These results suggest that the ecological impacts of high frequency, low intensity fires are likely to be lower than is often predicted based on assumptions of homogeneous landscapes and uniform burning.'

Conclusion

Wisdom, if we have any, is that fire – neither too frequent nor too seldom – will produce the best outcomes for biodiversity in much of the southern eucalypt forests of Australia. The same is true for woodlands, heathlands and grasslands.

It is obvious that there are small areas of forest that need less fire, especially those forests containing rainforest elements. However, even here we need to question for how long those elements will remain protected from bushfire, given changes in climate. And there are forests where almost the only type of fire is high intensity bushfire owing to their dominance of particular landscape positions and climates (for example, the mountain ash and alpine ash forests of the Dividing Range). That still leaves millions of hectares of eucalypt forests where low intensity fire can be safely and effectively used to limit fuel loads, mitigate bushfire risk, maintain forest health and assist in the conservation of species.

In contrast to the 8% or so of public land that is burned annually for fuel reduction in south-western Western Australia, less than 2% of public land is burned annually for fuel reduction in the eastern states. Despite strong recommendations in 2008 for a three-fold increase in annual fuel reduction from its own Parliamentary Committee, the Victorian Government has maintained the annual target of 130 000 hectares, a bit more than 2% of public land. It will always be difficult to plan and implement fuel-reduction burning over large areas of forest. Even with the very large injection of funding and resources required, the sheer size of the forest estate and the vagaries of terrain and climate will ensure that irrespective of the targeted area (in hectares) to be burnt, much of it will remain unburnt.

We are, of course, more than aware that many people in the wider community support our views, and that many are totally opposed. A future of extensive and intensive bushfires like we have had in the past 7 years is no way to manage the land and its diversity. We regard these sort of bushfires in this day and age as unacceptable, and we disagree totally with Bradstock's (2008) statement that 'it is arguable that the "disaster" paradigm is inextricably linked with a command and control style of management aimed at elimination of the problem', and he goes on to note that that 'control over the fire "problem" was sought via control of fuel' and that there are limits to such an approach. The extreme of this attack on management is expressed by Paul Collins (2006):

'One of the most striking things about discussion of forest and bushfires is the kind of rhetoric that is often used . . . I am referring specifically to the assumed, apparently unconscious attitude that nature needs to be 'managed'. It is apparent in commonplace roadside signs erected by State forestry commission: 'Managing the State's Forests Sustainably' – as though forests couldn't manage themselves sustainably and needed a government department to sort them out . . . The mania to manage one's entire environment is a classical symptom of the human fear of loss of control, of an inability to stand back from nature and allow it to be itself, of a failure to show some humility toward a world that has been in evolution for millions of years and does not need us or our technology . . . A wildfire exposes the sheer impotence of humankind, and for the control-freaks among us that is almost unbearable'.

Campbell (2008) tries to take the middle ground (having previously cast doubts about the worth of fuel-reduction burning):

'In recent years, governments have invested hundreds of millions of dollars in trying to put out large bushfires, and insurance companies have paid out even more in claims. From a greenhouse perspective, it would be highly desirable for Australia to shift a bigger proportion of this expenditure into early fire detection and rapid response suppression. Similarly from an (natural resource management) perspective, the biodiversity, water yield and water quality advantages from keeping fires as small as possible are considerable.'

We agree: bushfires must be kept as small as possible. To achieve that, fire prevention and fire suppression go hand in hand, and fuel reduction is an essential part of fire prevention. 'Even an ecologically unsophisticated program of numerous small, low intensity "hazard reduction fires" designed specifically to reduce fuel loads (may be preferable) to a "bushfire disaster" mode of fire management' (Bowman 2003a).

To reduce the risk of bushfire, to maintain diversity and to manage soils, carbon and water, we should use prescribed fire more widely and with fewer restrictions than has been the case in the past. Fewer restrictions does not mean less wisely. It does mean setting aside poorly considered fire exclusion zones. Using fire in vast areas of forest, for example, that stretch eastwards and uninterrupted from east of Melbourne to the Victorian border, requires an experienced workforce of people who understand fire. A workforce that understands local conditions, local weather and local topography. A workforce that understands that low intensity, fuel-reducing fires will be patchy. A workforce that understands there will be different prescriptions, depending on forest type, and that these must be sufficiently flexible to allow for changes in climate that will not be uniform across the landscape. Seasonality of burning is no longer what it once was. Changing climates has seen to that. Above all, we need a workforce that is well-trained in the science of fires and forests, and we need this workforce to be based in country towns – where the action is – and not aggregated in city office blocks.

'There is no alternative to the planned use of fire. Authorities have invested heavily in fire suppression plans and equipment (water bombers, fire tankers, bulldozers and so forth), while not meeting even modest targets for prescribed burning. However, even the best equipment we have for fire suppression is of limited value in the face of large accumulations of fuel. The argument is therefore somewhat circular. Unless we burn the bush in a controlled way, it is inevitable that the bush will burn, uncontrolled and leaving its legacy of death and destruction that will simply add strength to the view that 'all fire is bad' (Attiwill and Wilson 2006).

Our conclusions are not uniquely Australian, nor are they new. At the global scale, two most eminent fire ecologists (Bond and van Wilgen 1996) conclude their chapter on fire and management:

'Fire is an integral part of the dynamics of the Earth's surface, and the world's biota have evolved to cope with the phenomenon of recurrent fires. Fire is so essential to the maintenance and manipulation of ecosystems that it has to be considered and used by all managers of such ecosystems. Despite the concerns about the effects of fires on the dynamics of the atmosphere, there can really be no serious consideration of the elimination of fire from major portions of the Earth's surface to alleviate these concerns, Fire is necessary for the maintenance of biological diversity and the production of livestock. It is, in any case, an inevitable consequence of a combination of fuel, weather and sources of ignition. Managers who have attempted to exclude fire from ecosystems in the past have learnt the folly of their actions the hard way.'

Appendix: Ecological effects of repeated low-intensity fire in mixed eucalypt foothill forest in south-eastern Australia

We include here two unedited extracts from: Department of Sustainability and Environment (2003) Ecological effects of repeated low-intensity fire in mixed eucalypt foothill forest in south-eastern Australia: Summary report (1984–1999). Fire Research Report No. 57. Department of Sustainability and Environment, Victoria.

Extract 1

The 2003 series of Fire Research Reports developed from the Wombat fire effects study are:

No.	Title
57.	Ecological effects of repeated low-intensity fire in a mixed eucalypt foothill forest in south-eastern Australia – Summary report (1984–1999) – Department of Sustainability and Environment
58.	Effects of repeated low-intensity fire on the understorey of a mixed eucalypt foothill forest in south-eastern Australia – K.G. Tolhurst
59.	Effects of repeated low-intensity fire on fuel dynamics in a mixed eucalypt foothill forest in south-eastern Australia – K.G. Tolhurst & N. Kelly
60.	Effects of repeated low-intensity fire on carbon, nitrogen and phosphorus in the soils of a mixed eucalypt foothill forest in south-eastern Australia – P. Hopmans
61.	Effects of repeated low-intensity fire on the invertebrates of a mixed eucalypt foothill forest in south-eastern Australia – N. Collett & F. Neumann
62.	Effects of repeated low-intensity fire on bird abundance in a mixed eucalypt foothill forest in south-eastern Australia – R. Loyn, R. Cunningham & C. Donnelly
63.	Effects of repeated low-intensity fire on terrestrial mammal populations of a mixed eucalypt foothill forest in south-eastern Australia – M. Irvin, M. Westbrooke & M. Gibson
64.	Effects of repeated low-intensity fire on insectivorous bat populations of a mixed eucalypt foothill forest in south-eastern Australia – M. Irvin, P. Prevett & M. Westbrooke
65.	Effects of repeated low-intensity fire on reptile populations of a mixed eucalypt foothill forest in south-eastern Australia – M. Irvin, M. Westbrooke & M. Gibson
66.	Effects of repeated low-intensity fire on tree growth and bark in a mixed eucalypt foothill forest in south-eastern Australia – K. Chatto, T. Bell & J. Kellas

Extract 2

The following is a summary of some of the major findings of this study.

Forest fuel dynamics

- Surface fine fuels in this forest have an average steady state level of 16 tonnes per hectare, but seasonal variations may result in the fuel loads ranging from 9 tonnes per hectare to 26 tonnes per hectare.
- Surface fine fuel reaccumulates to within 90% of the long-unburnt state within four years of either spring or autumn burning at the fire intensities studied. The impact of prescribed burning on surface fuels is therefore relatively short-lived. (While the study reported here focused on surface fuels, other studies, which have focused on the overall fuel complex, have shown that the effects of prescribed burning may last 15–25 years.)
- The rate of reaccumulation of surface fine fuel is not significantly affected by the season of burn or the burn frequency.

Vegetation

- Over a 14-year period, no plant species were either lost or gained as a result of up to four successive spring fires or three successive autumn fires. The relative cover/abundance increased for about 30% and decreased for about 20% of the species.
- Short-rotation spring burning favoured Austral Bracken (*Pteridium esculentum*), herbs, geophytes and Poa (*Poa sieberiana*), and disadvantaged Forest Wire-grass (*Tetrarrhena juncea*), rushes, shrubs, small shrubs, legumes, trees, small trees and climbers. Long rotation spring burning favoured legumes and, to some extent, Poa.
- Short-rotation autumn burning favoured only herbs, but disadvantaged Bracken, Wire-grass, Poa, rushes, legumes, small shrubs, trees and geophytes. Long-rotation autumn burning favoured Poa, rushes, legumes and trees by providing a regeneration opportunity followed by a growth period. Bracken, shrubs and small shrubs were the only plant groups that seemed to be disadvantaged by long-rotation autumn burning.
- There were subtle changes in the forest understorey in the absence of fire. These changes were only small on a year-to-year basis, but amount to significant changes over a period of a decade or more.

Invertebrates

- Neumann and Tolhurst (1991) found that no long-term changes in activity or abundance among invertebrates occurred following a single low-intensity prescribed burn in spring or autumn; Collembola and Diptera populations recovered to pre-burn levels within one year, and earthworms within three years
- Three low-intensity prescribed spring burns within eight years had minimal impact on litter arthropods in foothill forest, although the effect of short-term rotational burns on the abundance of the Coleoptera and Diptera is less clear due to the significant change in activity levels also observed in the Control Treatment Areas.

Bats

- Bats range at least 1–5 kilometres while foraging. Therefore, to study bat ecology on a landscape scale, a much larger area than the FESA Treatment Areas is necessary.
- The information gained from this research program was insufficient to predict the effects of repeated fuel-reduction burning on bats. However, to date, the data suggests no effect in the differing Treatment Areas.

Reptiles

- None of the reptile species studied was favoured by a particular given burning treatment.
- Unburnt microhabitats (particularly logs, deep beds of leaf litter and areas frequently missed by low-intensity fire, such as gullies) provide important refuges and food, shelter and oviposition sites in the post-fire period. While some reptiles able to survive fire utilise alternative microhabitats, the rapid recovery of preferred microhabitat components is of major importance in the post-fire survival and recovery of reptiles.

Terrestrial mammals

- No particular burning treatment favoured either of the species studied, but habitat preferences were observed. Although *Antechinus agilis* (Brown Antechinus) and *Rattus fuscipes* (Bush Rat) have different habitat preferences, the survival and recovery of both species depend largely upon retention of unburnt habitat patches.
- Due to the timing of breeding and juvenile recruitment of *A. agilis*, autumn burns were found to have a greater impact on the population than spring burns. Population of *A. agilis* required at least 16 to 24 months to recover to pre-fire levels after autumn burns.
- Populations of *A. agilis* were significantly higher two to three years following spring and autumn burns than in long-unburnt areas.
- Populations of *R. fuscipes* took three breeding seasons (36 months) to recover when more than half of its preferred habitat was burnt during spring, but there was no recovery during the same period when the entire habitat was burnt.
- A single spring fire produced a greater impact than a single autumn fire on *R. fuscipes* populations in the first 12 months post-fire.

Birds

- The effects of burning in spring or autumn are quite similar (on subsequent bird abundance compared with unburnt areas) and there is no clear case for burning in one season and not the other. However, the beneficial effects of fire appear to be somewhat greater with autumn burns (probably because they are more intense) and the detrimental effects somewhat greater with spring burns (probably because birds are nesting at that time).
- Some birds respond positively to fire, and some species may depend on it in this forest type – e.g. White-winged Chough (*Corcorax melanorhamphos*), Spotted Quail-thrush (*Cinclosom punctatum*), Blue-winged Parrot (*Neophema chrysostoma*) and Red-browed Finch (*Neochmia temporalis*). Fuel-reduction burning serves to provide ephemeral patches of bare-ground habitat at the landscape scale, though it does not mimic the patterns expected under a regime of occasional extensive wildfires.
- Birds may also need access to unburnt vegetation within their home ranges, especially in the immediate aftermath of a fire as noted for mammals by Newsome et al. (1975), Catling and Newsome (1981), Humphries (1994), Tolhurst (1996a) and Friend (1999). The present study provides no information on the ability of birds to persist in areas subject to such treatment.

Soils: carbon, nitrogen and phosphorus

- There was a significant decline in carbon and nitrogen in surface soils due to repeated low intensity fires at three-yearly intervals. Furthermore, there was some evidence of a change in the quality of organic matter. However, there was little, if any, change in carbon and nitrogen from less-frequent fires (ten-year intervals), indicating that this strategy can be expected to maintain soil organic matter in the long term.

- Observed changes in extractable phosphorus were not attributable to fire.
- In the long term, low-intensity fires occurring in these foothill forests at less than a ten year frequency can be expected to lead to a decline in soil organic matter and soil fertility.

Tree growth, mortality and bark thickness

- Bark loss due to burning was shown to depend significantly on burning treatment. Both season and frequency of fires were important in the loss and recovery of bark after fire. Bark loss was found to be greater as a result of autumn burning than spring burning. This pattern was found to be strongly related to Soil Dryness Index and poorly related to fire intensity.
- Tree mortality rates during the study were relatively low and not attributable to any one particular treatment. There was considerable variation between FESAs, with fewer deaths recorded for areas with relatively 'cooler' burns (e.g. Blakeville) and more deaths at Treatment Areas with higher-intensity burns that often resulted in crown scorch (e.g. Kangaroo Creek).

Concluding comments

The Wombat Fire Effects Study has demonstrated that some of the conclusions drawn from short-term studies or space-for-time studies can be misleading. Given the longevity of a forest ecosystem, studies undertaken in a period of less than ten years may erroneously attribute trends or variations to unlinked causes, or overlook underlying trends altogether. Longitudinal studies, such as this one, provide a better insight into the dynamics of a forest ecosystem.

The references cited in these two extracts are included in References at the end of this book.

9

Concluding comments: fuel reduction is essential for effective fire management in Australia

10 key points: the case for fuel-reduction burning

We summarise the themes that we have developed in this book as **10 key points**:

1. **Fire is a natural element of Australia's ecology.** Australian ecosystems – the bush – have evolved with fire and are in many ways dependent on fire for regeneration and for the rejuvenation of ecological processes.
2. **The nature of fire in Australia has changed over the millennia.** Large-scale, lightning-caused fires were followed by finer-scale mosaic burning after Aboriginal settlement some 45 000–70 000 years ago. Following European settlement, larger and more intense fires ('feral fires', because of increased fuel loads and shifts in seasonal timing and frequency) have obliterated the pre-existing habitat mosaic created by Aboriginal landscape burning.
3. **This change in fire regime in conjunction with the introduction of cattle, sheep and other domesticated and feral herbivores, has caused the decline, and in some cases the extinction, of many mammal and bird species.** 'The transition from traditional Aboriginal to European fire management is a major ecological and evolutionary event that, while being different in character, is of the same significance as the Pleistocene colonisation of Australia by the ancestors of Aboriginal people' (Bowman 2003a).
4. **Given the right conditions, the Australian bush is highly flammable.** Under extreme conditions, bushfire can be so intense that no fire-fighting capabilities of any nation could stop them.
5. **We will not be able to eliminate bushfires**, whether in state forests, national parks or wilderness; given that we have forests, then we will have forest fires, and this stark reality is true, not only for our forests in Australia, but for forests over much of the world.
6. **The occurrence of wide-spread, high-intensity fires will not reduce unless we recognise the need for intensive fuel-reduction programs.** The amount of fuel is the only part of the fire triangle – ignition, fuel and weather – that we can control. With climates predicted to become increasingly hotter and drier, the case for immediate action is further strengthened (Box 9.1).
7. **The evidence for the requirement for planned management of fire (including fuel-reduction burning and lightning strikes) is overwhelming, not just in Australia but worldwide.**
8. **The technology is now available to manage fire in a way that maintains biodiversity and markedly reduces the hazard to human life and property.** The only barriers to the implementation of this technology are the provision of sufficient funding for fire management and the removal of the ideological barriers to fuel-reduction burning.

9. **We must now develop programs of fuel-reduction burning over all tenures** (national parks, reserves, wilderness areas, state forests and so forth) that are aimed not just at reducing fire magnitudes and intensities, but aim at fire regimes that include the combination of fuel-reduction burns and natural fires so that ecological diversity is maintained, if not enhanced, while the accumulation of fuels is decreased. Management *must* have well-defined aims for diversity (controlled burning in Kakadu, elsewhere in Australia's north and in many ecosystems, such as the diverse ecosystems of south-west Western Australia and the heathlands at Wilson's Promontory in Victoria, provide excellent examples for us to follow).

10. **Properly managed fuel-reduction burning causes far less damage to life and property and to diversity and ecological processes than intense and uncontrollable bushfire.** The reintroduction of large-scale fuel-reduction burning programs for the maintenance of biodiversity and for the protection of life and property demands an intensive effort by government agencies, and a major program to stimulate public awareness.

Box 9.1:   'Are Big Fires Inevitable?'

This is an extract from a larger article by Attiwill and Packham (2009).

Various management authorities are telling us increasingly that big fires are inevitable, largely due to global warming. Global warming is, however, only one part of the story.

'Are Big Fires Inevitable?' is the title of a national bushfire forum convened by the Bushfire CRC in Canberra, February 2007 . The lead paper at this forum was given by Jerry Williams, former National Director of Fire Management, US Forest Service, and was titled 'The Megafire Reality: Redirecting protection strategies in fire-prone ecosystems'.

While Williams' paper has an emphasis on the United States, its relevance to fire-prone ecosystems around the world is clear. Conventional strategies and conventional doctrine have been to increase fire fighting capacity and suppression force in line with increasing threat of bushfire. Despite the fact that limits to the effectiveness of suppression are well known, we have continued to concentrate on building our suppression force. To a very large extent, suppression has been effective, except for the few fires that have become uncontrollable and grown to become megafires.

A view from the United States

Williams puts the problem in the United States into perspective:

- In the past 15 years, there has been a three-fold increase in the fire-fighting budget. Nevertheless, the area burnt is increasing, and fire extent and severity is the worst in history. 'Megafires are not occurring due to a want in funding. The worst (bushfires) on record are coinciding with the highest preparedness budgets we've ever seen appropriated'.
- In 2006, suppression costs on federal lands alone exceeded US$1.9 billion, yet more than 3.9 million hectares were burned by bushfire.
- 'The fire protection strategy in America is expensive and promises to become more expensive . . . It has gotten to the point where – at this level of suppression

spending – other work for wildlife, watersheds, forest health, and recreation is going undone. Some are beginning to fear that we are headed to an awful predicament where the cost of emergency response continues to grow without discernible progress, but, worse, reaches the point where there is not enough money remaining to ever hope of mitigating the underlying problem.'

- Of the 10 000 or so fires that the U.S. Forest Service deals with each year, close to 99% are contained at the initial attack phase. The 1% of fires that escape account for 95% of the total area burned and about 85% of the total suppression costs.
- Before 1987, bushfires greater than 2000 hectares were relatively rare. Since 1998, there have been more than 200 fires greater than 20 000 hectares.

Williams argues that the increase in megafires is a powerful indicator that the fire protection strategies we have developed are not working. We are facing three major challenges: global warming, over-accumulation of forest fuels, and growth of populations in the interface between urban and forest areas (that is, more and more people wanting to live in the bush). In meeting this challenge, should we reinforce current tactics or should we redefine strategies?

Williams' view is that the megafires in America are 'emerging as more of a land management issue than the more commonly perceived fire issue'. Forests that have been protected by minimum disturbance and by fire suppression over the years have built up huge fuel-loads that are now, together with increasingly hotter conditions, fuelling the hottest fires. The very values that we set out to protect – species habitat, catchments for water supply, and many others – are being destroyed. 'Ironically', says Williams, 'we may be inadvertently managing for high-consequence wildfires'.

How are we to deal with these increased fuel loads throughout our forests? Techniques and management of prescribed burning to maintain low fuel-loads are well developed, but in the past 30 or 40 years, public opposition to the use of fire to control fire has intensified, nowhere more so than in the increasing area of national parks throughout the world. As a society, we began to believe that doing nothing – letting Nature look after itself – was an appropriate way to mange the land. And so, according to Williams, there is an implicit bias in many of the laws that govern land management. For example, prescribed burning must comply with the Clean Air Act, the National Environment Policy Act, and many others such as the Endangered Species Act. However, Williams notes that bushfires 'remain largely exempt from these requirements, even though their consequences may be much more severe'.

Williams concluded that changes in climate, fuel-loads and the urban/bush interface present real and serious threats. Leaders must address three questions:
- 'Are big, dangerous, destructive and costly bushfires inevitable?
- Can we see these trends and act on them?
- Can we re-direct (bushfire) protection strategies?'

At the time of writing (23 October 2007) fires are raging through southern California. A state of emergency was declared as fire ripped through the Malibu area, studded with homes of the rich and famous, living in the urban/bush interface. All of this, despite the fact that California has the largest fire protection capacity of any place in the world, with a budget for fire preparedness of more than US$3 billion (from Williams).

10 key reasons (or excuses) for inadequate programs of fuel-reduction burning

The management of fuel-reduction burning within the knowledge of fire behaviour is the legacy in Australia of the fire champions of the Forestry and Timber Bureau in Canberra, Harry Luke and Alan McArthur – work that started in the 1950s and continued under Phil Cheney's direction when the Bureau was incorporated within the Division of Forestry and Forest Products, CSIRO.

At the same time as Luke and McArthur were researching fire behaviour and control in Australia, Harold Biswell was doing similar work at the University of California at Berkeley. Biswell was a powerful advocate for fuel reduction, and this was a revolution – a paradigm shift – in fire management in the western United States. Biswell faced the same opposition to fuel reduction as we are seeing today in Australia, and he lists a number of reasons (or excuses, as he called them) for not controlling fuel loads in his seminal book (Biswell 1989).

We have built on and amended Biswell's list within an Australian context.

1. **The idea that all fires are bad**. This is a long-standing and firmly-entrenched view. It is wholly understandable, given the frighteningly destructive power of a bushfire; given that experience, it is difficult to view fire as good. However, the forest manager is faced with the day-to-day task of having to control *unplanned fires* including natural fires caused by lightning that may threaten diversity, other forest assets (water, topsoil, streams and rivers) and assets on adjoining lands. On the other hand, the forest manager must plan for 'good' fires – *planned fires* lit to achieve specified outcomes such as the reduction of fuel loads that have increased as a consequence of controlling natural fires and fires lit to maintain the health and diversity of forest ecosystems or for regeneration of disturbed or degraded ecosystems (Hodgson 2004).

2. **So-called 'controlled' fires can get away**. Of course they can, and the media revels in it. We should be critical if a fire gets away due to poor management or poor planning (for example, lit at the wrong time of the year or left, not properly extinguished, or when it is a danger to assets). We are loud in our praise of our fire-fighters when they tackle bushfires of hundreds of thousands of hectares, but we are critical of them when a fuel-reduction fire, planned to reduce the intensity and spread of bushfires, covers tens or hundreds of hectares more than was planned. The risks of fuel-reduction burning are minimal, given proper planning and a well-trained and experienced crew. We agree with Biswell: 'Isn't it more dangerous to live with fire hazards through periods of low humidity and high winds in the late summer than to burn them out under prescribed conditions?'

3. **Fires produce smoke, and we don't want that**. People in the cities, especially those with respiratory problems, do not like the smoke generated by fuel-reduction burns in autumn; there is enough pollution in the cities already. Winemakers may have to battle against 'smoke-taint' in their wines. These are difficult problems. They are problems that, in one of the most fire-prone areas on Earth, we have to learn to live with. If we do not burn to reduce the fuels, then fuel loads will steadily increase to 20 or more tonnes per hectare. When these fuels burn in an uncontrolled bushfire, as they inevitably will, the smoke will be prolonged. For example, the 2002–2003 alpine fires and the 2007 Great Divide fires in Victoria each burnt for about 60 days. The Kilmore-Murrindindi fires in Victoria, on Black Saturday 7 February 2009, extended south in to the Yarra Valley fire as the grapes were ripening. Because smoke-taint in wine is at least partly due to absorption into the skins of grapes around about this critical time (veraison – the transition from growth to ripening and change of colour) it is clearly of benefit to have cooler, limited burns in autumn when the grapes have been harvested than intensive and widespread fires in late summer.

Box 9.2:   Grapes versus fuel reduction: a judgement from Western Australia

A judgement[30] handed down by Justice Graeme Murphy in the Supreme Court of Western Australia on 12 March 2010 is of great interest and we have noted several of its features elsewhere. The plaintiffs were a group of winegrowers whose grapes had been tainted by smoke from a controlled burn carried out near Pemberton, Western Australia by the Department of Conservation and Land Management (now called Department of Conservation and Environment) in March–April 2004. The fact that smoke from the fire had tainted the grapes was not at issue; the issue was whether the Department owed a duty of care to the winegrowers.

As we have mentioned previously, Justice Murphy noted that an assembled group of experts had agreed that prescribed burning is a necessary part of public land management to meet biodiversity, economic, silviculture, water and community protection objectives, and that prescribed burning is the only effective way of reducing potential impacts of wildfires at a landscape level.

However, Justice Murphy went on to quote from the relevant Forest Management Plan that proposes actions at the whole of forest and landscape scale for the purpose of using and responding to fire in a manner that:

- *'optimises the maintenance of forest ecosystem health and vitality*
- *promotes the conservation of biodiversity*
- *controls adverse impacts of fire on the social, cultural and economic values of land managed by the Department and adjoining land*
- *minimises the risk of smoke emanating from prescribed burns impacting on population centres and other sensitive areas.'*

Among the actions proposed, the Forest Management Plan states that the Department will:
- maintain a competent fire management, suppression and response capability
- prepare and maintain a fire management plan and smoke management guidelines
- undertake an annual prescribed burning program in a manner that:
  a)   is in accordance with the fire management plan
  b)   is in accordance with the smoke management guidelines
  c)   has regard to the Goals for Understorey Structural Diversity
  d)   considers any special vulnerability of fauna and flora known to exist in a particular area to burning in that area
- consult with stakeholders and interested community members in a manner that seeks to develop community understanding of and support for, and enable constructive discussions and deliberations on, the planning and implementation of prescribed burning and other fire management programs.

Justice Graeme Murphy found that the case brought by the winegrowers was inconsistent with the statutory functions of the Department – that is, to carry out fuel-reduction burning 'so as to serve the overall objectives prescribed by the Forest Management Plan' – maintaining forest health, conserving biodiversity, controlling adverse effects of bushfires. Justice Murphy concluded that 'that a duty of care, as alleged by the plaintiffs, or otherwise, did not exist. If such a duty did exist, I find that it was not breached. Accordingly, the plaintiffs' action must be dismissed'.

However, atmospheric conditions in autumn are often quite stable so that the smoke is not quickly dispersed (Box 9.2).

4. **We need more research (see also Box 9.5).** We have worked in universities all our lives, and so we would never speak out against research. Research that is of high quality, and research that is innovative and needed. However, the general argument that we should not allow repeated burning to reduce fuels because we have not studied the effects on all of the biota has no scientific basis. Anecdotally, we have the ecosystems that we have because of repeated burning over millennia. Scientifically, we have many studies of key species from which we can formulate programs of fuel reduction. As well as getting on with the business of fuel reduction, we must ensure that future research is directed to key understanding in the areas of diversity and ecological processes. We should also ensure that research tackles the important question: what is the effect of excluding fire from Australian ecosystems? In the first decade of the 21st century, fire management has become disaster management. 'There can be no doubt that some styles of management are more sympathetic to biodiversity and ecosystem services than others. I believe the currently ascendant 'bushfire disaster' mode of management is ultimately more destructive of biodiversity than a program of recurrent fires to reduce fuel loads' (Bowman 2003b).

5. **There aren't enough days suitable for fuel-reduction burning.** The number of days available for fuel-reduction burning is not a given. It is determined by the rules and prescriptions and limits that are set by the management agency. Given the accountability that rests with the manager, and the knowledge that if something goes wrong the buck stops with the manager (with little or no support from the government in power), it is inevitable that the rules and prescriptions and limits will be highly conservative, and that it will certainly be true that 'There aren't enough days suitable for fuel-reduction burning'. In fact, there are so few days suitable for prescribed burning under the current prescriptions that even the small (about 2% of State land) target areas for fuel-reduction burning set by the Department of Sustainability and Environment, Victoria, have not been met. Over the years 2000–2008, the actual area burnt was only 95 262 hectares per year, or 78.6% of the target area. Fuel-reduction burning demands local knowledge and experience. The increasing centralisation of management makes choosing a good day for burning increasingly difficult. A second factor at play here is the trend established through single-species approaches to setting fire-return intervals. That trend results in ever-smaller areas to be treated by any given fuel-reduction fire, and thus much reduced availability of resources (people, trucks and equipment) on the days when fires can be set. 'The Peoples Review' (Attiwill *et al.* 2009) following the 2002–2003 and 2007 bushfires in Victoria encouraged comments from the people to see how we can improve our efforts, not just in fire suppression but particularly in fire prevention. The overwhelming view was to return fire management, including fuel-reduction burning, to the people living and working in rural Victoria; we include some recommendations from this Review (Box 9.3).

6. **Burning makes ecosystems more flammable.** We are not sure where, how and why this notion originated. It seems to be a claim coming from those who oppose any sort of management of, or interference with, the bush (apart from demanding that lightning strikes are dealt with promptly). The general idea seems to be that interference by fuel reduction, logging, and so on creates a drier, more sclerophyllous understorey. As far as we are aware, the evidence in the scientific literature for such a change is very limited. The proponents use, among other things, Judge Stretton's (1939) report: 'The fire stimulated grass growth; but it encouraged scrub growth far more.' As we have previously discussed, Stretton had very limited published science to draw upon and his comment on this is at odds with all of the evidence of the nature of forests and woodlands at the time of European settlement.

Box 9.3:   Some recommendations coming from 'The Peoples' Review' (Attiwill *et al.* 2009) on the need for local experience, knowledge and expertise in all aspects of fire management

1.2.6 People's future involvement – People's Fire Forum, Fire Policy Group, and local control of prescribed burning

In many fields of endeavour, the main location of knowledge is in institutions created to maintain and develop that knowledge. In the case of fire, however, much of the basic knowledge about local topography, fire and wind conditions lies not with city-based institutions, but with the local people themselves. This is why the people need such a strong voice. The following recommendation is not directed to government but to the people of Victoria – to urge them to pick up where this People's Review concludes and carry forward the imperative that the people have a right to be heard, their views assessed and changes implemented.

Recommendation 9: The People's Review recommends the establishment of a State-wide peak fire forum for the people, which we shall call the People's Fire Forum.

The development of fire prevention and suppression policy must be in the hands of people of all ecological persuasions. There is undeniably strong anecdotal evidence that the staff of Parks Victoria and Department of Sustainability and Environment is, simply put, seen as being too green. Fire policy must broaden from bureaucratic control and city-based politics to reflect the views and aspirations of the community, especially of rural communities. After all, it is they who live in the area, respond as volunteers to fight fires and bear massive social and economic cost of bushfires.

Recommendation 10: The People's Review recommends that fire prevention and suppression policy be set and reviewed by a Fire Policy Group that includes representatives of the People's Fire Forum.

Recommendation 11: The People's Review recommends that Prescribed Burning Groups, on which local communities have at least 40 percent of the representation, decide arrangements for prescribed burning including targets, timing, location and accounting.

An associated claim is that **old-age forests are more fire resistant than younger forests**. This claim mostly refers to the wetter forests such as karri in the west and mountain ash and alpine ash in the south-east. Again, we are not aware of any evidence in the scientific literature in support of increasing fire resistance with age. Indeed, one of the ecological tragedies of the Black Saturday 2009 bushfires in Victoria was the complete killing of old-age (at least 350 years) mountain ash forest in the Wallaby Creek catchment and an almost 100% killing of similar old-age forests in the O'Shannassy catchment.

7. **We don't know enough about Aboriginal burning to have an equivalent program of fuel reduction.** Aboriginal burning of the dry sclerophyll forests and woodlands was constant. Even if we had exact knowledge of Aboriginal burning, a firestick approach is obviously inappropriate given the vastly changed patterns of occupation and settlement.

8. **Fuel-reduction burning is too costly; there is not enough money.** Whether or not there is enough money is entirely a matter of priorities. Governments around Australia have spent a great deal of money on building up the capability for fire suppression, and we do not

question the validity of that. However, the cost of fire suppression in southern Australia over the past 8 years has been enormous. In Victoria, most of the money for fuel reduction has been spent on asset protection in specific areas. In comparison, the money needed per hectare of broader-scale fuel reduction is much less than that needed in sensitive areas of asset protection. After all, it is at the landscape level that fuel loads promote the sort of crown fires that end up in a firestorm. We ask the question: 'how should governments provide funding for the landscape-scale fuel-reduction burning that is required to adequately reduce risks, free from year-to-year political pressure?' (see Box 9.4).

9. **It's all a matter of politics.** Spending money on water bombers, tankers, communications and so forth for better fire-fighting capability results in tangible and accountable assets. Spending money on fuel reduction results in good figures on a broadsheet, but it may anger city people (because of smoke) and those among the conservation and other groups who

Box 9.4:   Fire, money and politics

Fire has become a political issue in recent years and is set to become more so unless significant changes are made in overall approaches to fire, especially fuel-reduction fires. The involvement of politicians and senior bureaucrats in the issues surrounding the fighting of bushfires has been in plain view.

Over the past 20–30 years, governments have sought to move costs away from the public purse where they can be legitimately passed to private individuals. This is the so-called 'user pays' principle and is enshrined in many modern policy decisions. Examples include health insurance, toll roads and bridges, and user charges for national parks. Sometimes the private sector collects payments directly, and sometimes governments collect them and pass them on to private sector contractors for the service provided.

Another example is the fire levy that people pay additional to their house insurance. This levy is used to fund fire-fighting services – both the professional metropolitan brigades and the volunteer rural brigades. There is considerable logic behind such a levy. First, it is attached to the principle of insurance – spreading the risk based on the assumption that (hopefully) not everyone suffers loss at the same time. Secondly, the greater the value of the property insured, the greater the levy and the more money that flows to fire-fighting agencies. State and federal governments also contribute to the costs of running, and especially equipping (trucks, helicopters and planes), the fire-fighting agencies, but the under-pinning by insurance levies is now an accepted and crucial part of their balance sheets. The growth in budget of volunteer-based fire-fighting organisations is testament to the success of the policy, even though it does little to ensure we have enough volunteers owing to the 'urbanisation' of the population.

However, the 'insurance and levy' approach to risk mitigation via fire-fighting stands in contrast to the equally valid and just-as-important mitigation of risks via fuel-reduction burning. For the latter, all of the costs have remained with the public purse. Consequently, there is no financial underpinning of the costs of fuel-reduction burning. Instead, each year bureaucrats from land management agencies responsible for fuel-reduction burning have to request funds from Treasury to pay for their annual program. Naturally enough, this leads to a situation where Treasury officials have to weigh up the importance of fuel-reduction burning relative to the importance of funding for hospitals, police, schools and a myriad of other competing and important community needs. Not surprisingly, fuel-reduction burning is not always high on the list of priorities, and falls in importance with the passage of time since the last major bushfire; a reduced importance that is all too easily supported by those opposed to fuel reduction on political or ideological grounds. We might compare and contrast the large infrastructure and budgets of the specialist fire-fighting agencies with now obvious lack in rural areas, of land management infrastructure, people and actions. Policies of 'benign neglect' seem more favoured, with governments only springing into action when there is a direct threat.

Two 'rules-of-thumb' might help to de-politicise fuel-reduction burning: (1) Funding provided each year for landscape-scale, fuel-reduction burning should be a matter of public record, as should its expenditure; (2) Mechanisms for the provision of those funds should be transparent, free from political influence and proportional to the risks (especially fuel loads).

The second point carries with it some problems. Removing political influence of fuel-reduction burning is made more difficult by fire ignitions many kilometres away from where they ultimately destroy property and kill people. There is overwhelming temptation to focus on risks 'over the back fence', and to describe fuel reduction at that scale as 'strategic'. Individuals might be convinced they ought contribute to costs of reducing risks within their 'neighbourhood', but it will remain difficult to convince them they should contribute to costs associated with fuel-reduction efforts even 1 kilometre distant, let alone 10, or perhaps 100 kilometres away. However, unless fuel loads are reduced across the landscape, there is little hope of containing the worst fires once they have started and, under 'catastrophic' conditions, preventing them from becoming raging infernos capable of overwhelming the modest defences 'over the fence'.

are implacably opposed to interference with our natural resources. We seem to have a major contradiction. On the one hand, private landowners are urged every year to take the appropriate steps, and to spend appropriate money, to reduce their fire risk – a clear acknowledgement that fuel reduction is (among other preventative measures) essential to reducing the risk of fire. On the other hand, government landowners have allowed fuel loads to build up to the extent that, given the right conditions, no amount of fire-fighting equipment will be successful in fire suppression. It is all a matter of politics (see Box 9.4).

10. **Let it be an act of God.** We have previously cited the following piece from Paul Collins, broadcaster, writer, and previously a priest in the Catholic Church:

'One of the most striking things about discussion of forest and bushfires is the kind of rhetoric that is often used . . . I am referring specifically to the assumed, apparently unconscious attitude that nature needs to be "managed". It is apparent in commonplace roadside signs erected by State forestry commission: "Managing the State's Forests Sustainably"—as though forests couldn't manage themselves sustainably and needed a government department to sort them out . . . The mania to manage one's entire environment is a classical symptom of the human fear of loss of control, of an inability to stand back from

nature and allow it to be itself, of a failure to show some humility toward a world that has been in evolution for millions of years and does not need us or our technology . . . A wildfire exposes the sheer impotence of humankind, and for the control-freaks among us that is almost unbearable'.

We cannot countenance Collins' view of humans as being apart from nature. We have cleared large areas of land to grow food, and we use extensive areas for grazing. We have introduced animals, both herbivores and carnivores, many species of which have become feral. We have introduced grasses, climbers, shrubs and trees, many of which have adapted superbly and are now widespread weeds. We have built dams and weirs and altered the courses of rivers and streams. We have extended settlement so that many people live surrounded by the bush. We have produced so much carbon dioxide by burning fossil fuels and by cutting and burning the tropical rainforests that we are making southern Australia hotter and drier. We have altered fire regimes. It is intolerable that we should now say: 'Good old Nature', you know best, you go and fix that lot!'

When we go walking through the forests of the Great Dividing Range, we are awed by their extent. For example, in the Central Highlands of Victoria, about 90% of the pre-1750 area of the wetter forest types and 99.9% of the pre-1750 area of cool-temperate rainforest is still there, contrary to scare-mongering views (Commonwealth of Australia and Victorian Governments 1998; Attiwill *et al.* 2001). In the south-west of Western Australia, 82% of the pre-1750 extent of karri forest in the region currently remains. Of the 13 jarrah forest types recognised in the South-West Forest region, five can be classed as more-heavily cleared (only 47% remaining) and six are less heavily cleared (53% remaining).

These forests are now our inheritance and they cannot be left to look after themselves. Nor can they be left to a bushfire regime of disaster management – wait for a disaster, and then throw our resources of people and water-bombers at it. That hasn't worked – surely we haven't already forgotten the feral fires of 2002–2003 and 2007 that burnt for 2 months, each costing hundreds of millions of dollars in fire suppression attempts, let alone the tragedy of 2009 – and it won't work in the future.

We are the current stewards and managers of the land. Aboriginal people managed fire and tamed it. Forest managers are setting the example in how to manage and tame fire in Western Australia. We have plenty of precedents to learn from. So, to finish with mixed clichés, let's bite the bullet and carry the torch. And to go back to 'The Peoples' Review' (Attiwill *et al.* 2009); the people living and working in the bush don't need all these lofty words and arguments – it's just good old, plain common sense. If you are living with a time-bomb of accumulated fuels, do something about it!

Education and research

A point we have made repeatedly is that we know enough, including knowledge of diversity, to get on with the job of using fire to reduce the risks of bushfires. We also say that we need new and better knowledge of a great many aspects of fire in our forests, including fire behaviour (largely physical science, see Chapter 3) and the role of fire in ecological sustainability (largely biological science, see Chapters 5–8). Our knowledge will steadily accumulate as a result of the practice of fuel-reduction burning. This accords closely with the view of the Royal Commission (see Epilogue).

We cannot say the same about education (and training) – education at all levels so that people can learn how to live safely with fire. This challenge encompasses aspects of education and training within schools, universities, technical colleges and fire and land management agencies. It also encompasses continual professional development courses and public education programs.

This is not a trivial task. The events of 2009, and before that of 2006–07 and 2003, emphasise just how poorly bushfire is understood in Australia by urban planning agencies, utility companies and the general public. Some sections of emergency service agencies and land management agencies still fail to acknowledge the primacy of fire to their core business. National parks agencies, for example, still fail to properly monitor (or even attempt to measure) the impacts of bushfires on biodiversity, yet all too frequently their staff rail against the supposed 'risks to biodiversity posed by fuel-reduction burning'.

The education challenge requires recognition that we have had decades of a shift in staffing of land management agencies away from those trained in forestry schools and toward those trained in 'environmental science'. Incredibly to us, there is now an embedded 'anti-fuel-reduction burning' culture in some state government institutions.

There is now no School of Forestry anywhere within Australia. CSIRO Forestry no longer exists. Those tertiary institutions retaining an interest in forests now often deal only cursorily with practical aspects of fighting forest fires and fuel-reduction burning – this must be reversed. The natural and social histories of fire in Australia and elsewhere in the world are increasingly well documented and should be a formal component of degree courses in relevant disciplines.

School education programs must be better informed about 'living with fire'. Students must learn that using fire to reduce fuels is the only way that people can reduce risks from bushfires – and not have that message confused with or suffused by untested ideologies about the 'risks of prescribed fire'. Students must learn about the dependence of eucalypt forests on fire and their co-evolution with fire. A school education program *must* be as focused on inner urban schools as on outer urban schools. Students from the inner suburbs grow up to become part of the out-migration phenomenon and too many have almost no appreciation of the dangers or the fire ecology of the forests to which they move.

Community programs too have suffered in recent years from confused messages. Government agencies, and agencies and organisations supported by governments, have delivered messages and materials that promulgate the view that prescribed fires pose serious risks to biodiversity, yet fail to even mention how many plants or animals are killed by bushfires. Little attention has been given to education about housing and planning control. Even today, radio programs openly discuss ways by which homebuilders can 'get around' fire risk assessments and their implied building costs ('get another assessor'). If we were all better educated about the basic elements of fire behaviour, fuels, and fire ecology and dependence of eucalypt forests, then perhaps there might be fewer poor decisions about buildings and behaviour during fires.

It is vital that emergency service personnel receive much better education and training in the broader aspects of 'living with fire', in addition to their traditional base of fire-fighting skills and equipment and policies and procedures. Understanding the forests – how well they grow and produce fuel in relation to soils, topography, aspect and climate – is vital to decision making in the face of bushfires. Understanding human psychology and likely reactions to bushfires is obviously equally vital.

All of this requires resources and commitment. Books and other teaching materials must be written and prepared, lecturers must be employed, teachers must be brought up to speed with current knowledge, curricula must be changed, policies and guidelines for practises must be revised and implemented, and staff must be trained and educated.

Stephen Pyne (2007) provides a nice summary: 'A fuel array, a habitat, a landscape – nature does not determine by which prism a place might be viewed; culture does. A serious fire scholarship would begin with that irreducible fact'.

Epilogue: The Final Report of the 2009 Victorian Bushfires Royal Commission

The Final Report of the 2009 Victorian Bushfires Royal Commission (The Hon. Bernard Teague AO [Chairperson], Ronald McLeod AM [Commissioner] and Susan Pascoe AM [Commissioner] was released on 31 July 2010. The following extract is taken from Final Report Summary, page 15:[31]

'LAND AND FUEL MANAGEMENT

Prescribed burning is one of the main tools for fire management on public land. It cannot prevent bushfire, but it decreases fuel loads and so reduces the spread and intensity of bushfires. By reducing the spread and intensity of bushfires, it also helps protect flora and fauna. Ironically, maintaining pristine forests untouched by fuel reduction can predispose those forests to greater destruction in the event of a bushfire.

About 7.7 million hectares of public land in Victoria is managed by DSE (Department of Sustainability and Environment). This area includes national parks, State forests and reserves, of which a large portion is forested and prone to bushfire. DSE burns only 1.7% (or 130 000 hectares) of this public land each year. This is well below the amount experts and previous inquiries have suggested is needed to reduce bushfire and environmental risks in the long term.

The Commission recognises that prescribed burning is risky, resource intensive, available only in limited time frames, and can temporarily have adverse effects on local communities (for example, reduced air quality). Nonetheless, it considers that the amount of prescribed burning occurring in Victoria is inadequate. It is concerned that the State has maintained a minimalist approach to prescribed burning despite recent official or independent reports and inquiries, all of which have recommended increasing the prescribed-burning program. The State has allowed the forests to continue accumulating excessive fuel loads, adding to the likelihood of more intense bushfires and thereby placing firefighters and communities at greater risk.

The Commission proposes that the State make a commitment to fund a long-term program of prescribed burning, with an annual rolling target of a minimum of 5% of public land each year, and that the State be held accountable for meeting this target. DSE should modify its Code of Practice for Fire Management on Public Land so that it is clear that protecting human life is given highest priority, and should report annually on prescribed-burning outcomes.

To ensure continuing environmental protection, the State needs to improve its understanding of the effects of different fire regimes on flora and fauna. The Commission proposes that DSE expand its data collection on the effects of prescribed burning and bushfire on biodiversity. Maintenance and extension of data collection on Victoria's flora and fauna assets has not been a high priority. It needs to be improved so that more informed and scientifically-based decision making can accompany the development of prescribed-burning regimes that meet conservation objectives as well as accommodating bushfire safety considerations'.

The Final Report contains 67 recommendations. The Government's interim responses[32] are grouped in three categories:

- 53 recommendations were accepted in principle
- six recommendations were accepted in principle, but with further consultation on various topics deemed to be necessary
- eight recommendations require further consultation.

Among the eight recommendations requiring further consultation are issues such as the provision of community refuges, buy-back of developments in areas of unacceptably high fire risk, and the replacement of existing single-wire cables with aerial bundled cabling or underground cabling.

Among the six recommendations accepted in principle but requiring further consultation is Recommendation 57: 'The State fund commit to implementing a long-term program of prescribed burning based on an annual rolling target of 5% minimum of public land'. Consultation is needed, the Premier says 'regarding implementation and scale-up'. 'Mr Brumby said that the Government would speak to stakeholders about . . . increasing fuel-reduction burning to an annual rolling target of 5% minimum of public land, recognising that fuel-reduction burning must increase, but that there are strong views held in different sectors of the community'.

That is all very well, but begs the question: who makes decisions about prescribed burning and on what basis? As the Commission noted: 'the State has maintained a minimalist approach to prescribed burning despite recent official or independent reports and inquiries, all of which have recommended increasing the prescribed-burning program. The State has allowed the forests to continue accumulating excessive fuel loads, adding to the likelihood of more intense bushfires and thereby placing firefighters and communities at greater risk'.

As we finalise this book (October 2010), the Victorian Environment and Climate Change Minister Gavin Jennings stated that we have exceeded our targets: 'we have exceeded the planned burn target of 130 000 hectare and far exceeded the 20 year average of about 90 000 hectares. Last season, despite unfavourable cooler and wetter weather, our fire crews achieved 146 106 hectares, which is well in excess of our previous 130 000 hectare target'.[33]

The Minister announced that $382.4 million of additional funding will enable planned burning to be scaled up to 275 000 hectares over the next 4 years, with an ultimate target of 385 000 hectares a year, in line with the recommendations from the Royal Commission.

We have presented our strong view, a view that we believe is not only scientifically rigorous but is good common sense. We can now only wait to see how the State places scientific rigour and fire management within its structure of land and forest management. Our view is that the State must continue to endorse and increase fuel-reduction burning, rather than depending on an emergency response of fire suppression.

Endnotes

1 Grattan M (1983) PM tries to polarize the issues. *The Age*, Melbourne, 16 February.
2 *The Age* (1983) 16 February.
3 From the Bureau of Meteorology website, http://www.bom.gov.au/lam/climate/levelthree/
   c20thc/storm7.htm:
   'Late on the morning of 8 February 1983 a strong, but dry, cold front began crossing
   Victoria, preceded by hot, gusty northerly winds. The loose topsoil in the Mallee and
   Wimmera was quickly picked up by the wind, and as the front moved east, the soil collected into a large cloud oriented along the line of a cool change. At Horsham, in western
   Victoria, raised dust could be seen by 11 am; by noon it had obscured the sky.
       In Melbourne, the temperature rose quickly as the north wind strengthened, and by
   2:35 pm it had reached 43.2°C, a record February maximum. A short time later, a spectacular reddish-brown cloud could be seen advancing on the city, reaching Melbourne just
   before 3 pm. It was accompanied by a rapid temperature drop, and a squally wind change
   strong enough to uproot trees and unroof about 50 houses. Visibility plunged to 100 metres,
   and according to witnesses 'everything went black' as the storm struck. The worst of the
   dust-storm was over by 4pm, when the wind-speed dropped rapidly.
       At its height, the dust-storm extended across the entire width of Victoria, and was many
   kilometres across. The dust-cloud was some 320 m deep when it struck Melbourne, but in
   other areas extended thousands of metres into the atmosphere. It was estimated that about
   50 000 tonnes of topsoil were stripped from the Mallee (approximately 1000 tonnes of it
   being dumped on the city), leaving the ground bare, and exacerbating the effects of the
   drought. Open water channels in the north-west were clogged with sand and dirt. The
   combined effect of drought and dust-storm inflicted damage on the land that, according to
   the then President of the Victorian Farmers and Graziers' Association, would take up to 10
   years and tens of millions of dollars to repair.'
4 Rawson RP, Billing PR and Duncan SF (1983) The 1982–83 forest fires in Victoria. *Australian Forestry* **46**, 163–172.
5 Rawson RP, Billing PR and Duncan SF (1983) The 1982–83 forest fires in Victoria. *Australian Forestry* **46**, 163–172.
6 Mountain Views Mail (1991) 11 March.
7 Rawson RP, Billing PR and Duncan SF (1983) The 1982–83 forest fires in Victoria. *Australian Forestry* **46**, 163–172.
8 The metric ton (or tonne) is 10^6 grams, or 1000 kilograms. The formal SI (le Système international d'unités) unit is Megagram, but we have used tonne here because it is still mostly
   used in the forestry and fire literature.
9 For a full description of fire danger ratings for Victoria, see: http://www.cfa.vic.gov.au/
   firesafety/bushfire/danger-warnings.htm
10 These figures were prepared using a variety of sources, including Luke and McArthur
   (1978), Cheney (1981), and Tolhurst and Cheney (1999).

11 Dexter's (2005) data on deaths and economic losses come from a report by MIRA Consultants Ltd to Department of Conservation and Natural Resources, Victoria, November, 1992.

12 Dixon and Barrett (2003) found that smoke is a key agent in promoting germination in more than 400 species from 82 genera of the most dominant families, especially in the major families of temperate Australia including Cyperaceae, Restionaceae, Dilleniaceae, Rutaceae, Myrtaceae, Proteaceae, Epacridaceae, Haemodoraceae and Poaceae. The study by Chambers and Attiwill (1994) is just one of many studies that have aimed to sort out this very difficult 'ash-bed' effect. Heating the soil changes chemical, physical and microbial properties of the soil, as do treatments such as sterilisation and soil preparation such as 'sun-baking' used in some traditional agriculture. The explanation of greatly enhanced seedling growth on soil over which fire has passed remains uncertain. The cause has been ascribed to microbial antagonisms that are diminished after fire, to specific fungal attack in the mature forest that is eliminated after fire, to exudates from the mature trees that may be allelopathic, and to increased availability of phosphorus and nitrogen caused by heat and by the addition of ash.

13 Biochar is formed by incomplete combustion of plant matter, and can be a by-product of the generation of power from plant biomass. Atmospheric CO_2 is taken up in photosynthesis by plants and the C is incorporated in plant matter. Biochar (and Pyr C) is rich in C and is relatively stable; the formation of biochar therefore sequesters atmospheric CO_2 in the long term.

14 Biomass was originally defined as the *mass* of living matter (as the name implies). It is now generally used to mean plant matter.

15 1 Gigatonne (Gt) = 10^9 tonnes, or 10^{15} grams.

16 Gross Primary Production (GPP) is the total amount of carbon dioxide (CO_2) fixed (a chemical reduction) by photosynthesis. A proportion of GPP is lost (a chemical oxidation) by plant respiration (R), so that Net Primary Production (NPP) = GPP – R. Thus NPP is that part of GPP that is not respired by plants and hence becomes available to all of the dependent heterotrophs of the ecosystem.

17 Association for Fire Ecology Position Paper, Adopted December 3 2009. 'The role of fire in managing long-term carbon stores'. http://fireecology.net/docs/AFE_2009_Position_Paper_Carbon.pdf

18 Water potential is a measure of the energy status of water. Transpiration is the movement of water from soil (higher energy status) to atmosphere (lower energy status). Water potential in plants is measured as the balancing pressure needed to force sap from the plant tissue. Water potential is a relative index, with the water potential of pure water set at 0. Water potential measured just before dawn assumes equilibrium between water in the soil and water in the plant.

19 ML = megalitre = 10^6 litres, or 1 million litres.
    GL = gigalitre = 10^9 litres, or 1 thousand million litres.

20 Net primary production (NPP) is that part of gross photosynthetic production that is not respired by the plant and hence becomes available for utilisation by all the dependent heterotrophs of the ecosystem.

21 http://www.ag.gov.au/ema/emadisasters.nsf

22 In 2006, the number of Fire Management Zones was reduced to four: Asset Protection Zone, Strategic Wildfire Management Zone, Ecological Management Zone, and Prescribed Burning Exclusion Zone (Department of Sustainability and Environment, Victoria, 2006.

23 http://burnseverity.cr.usgs.gov

24 http://landsat.gsfc.nasa.gov/

25 2009 Victorian Bushfires Royal Commission. *Land And Fuel Management: Planned Burning*. Submissions of Counsel Assisting. Submission 700.001.0001.

26 http://decisions.justice.wa.gov.au/supreme/supdcsn.nsf/PDFJudgments-WebVw/2010WAS C0045/$FILE/2010WASC0045.pdf

27 *Sydney Morning Herald*, 20 November 2009, article by Ben Cubby.

28 Lower threshold is the minimum period between fires.

29 *The Australian*, 16 January 2010.

30 http://decisions.justice.wa.gov.au/supreme/supdcsn.nsf/PDFJudgments-WebVw/2010WAS C0045/$FILE/2010WASC0045.pdf

31 http://www.royalcommission.vic.gov.au/finaldocuments/summary/PF/VBRC_ Summary_PF.pdf

32 Premier releases interim response to Bushfires Royal Commission Report, Monday 2 August 2010
http://www.premier.vic.gov.au/newsroom/11351.html

33 http://www.gavinjennings.org/pageGen.cgi?id=1524, 17 September 2010.

References

Abbott I (2003) Aboriginal fire regimes in south-west Western Australia: evidence from historical documents. In *Fire in Ecosystems of South-Western Australia: Impacts and Management.* (Eds I Abbott and N Burrows) pp. 119–146. Backhuys Publishers, Leiden, The Netherlands.

Abbott I and Burrows N (Eds) (2003) *Fire in Ecosystems of South-Western Australia: Impacts and Management.* Backhuys Publishers, Leiden, The Netherlands.

Aber JD, Goodale CL, Ollinger SV, Smith M-L, Magill AH, Martin ME and Stoddard JL (2003) Is nitrogen altering the nitrogen status of northeastern forests? *BioScience* **53**, 375–390.

Aber JD and Melillo JM (1991) *Terrestrial Ecosystems.* Saunders College Publishing, Holt, Rinehart and Winston Inc., Orlando, Florida.

Adams MA (1996) Distribution of eucalypts in Australian landscapes: landforms, soils, fire and nutrition. In *Nutrition of Eucalypts.* (Eds PM Attiwill and MA Adams) pp. 61–76. CSIRO Publishing, Melbourne.

Adams MA (2007) Nutrient cycling in forests and heathlands: an ecosystem perspective from the water-limited south. In *Nutrient Cycling in Terrestrial Ecosystems.* (Eds P Marschner and Z Rengel) pp. 333–360. Springer-Verlag, Berlin.

Adams MA and Attiwill PM (1984a) Role of *Acacia* spp. in nutrient balance and cycling in Adams *Eucalyptus regnans* F. Muell. forests. I. Temporal changes in biomass and nutrient content. *Australian Journal of Botany* **32**, 20–15.

Adams MA and Attiwill PM (1984b) Role of *Acacia* spp. in nutrient balance and cycling in regenerating *Eucalyptus regnans* F. Muell. forests. II. Field studies of acetylene reduction. *Australian Journal of Botany* **32**, 217–223.

Adams MA and Attiwill PM (1986) Nutrient cycling and nitrogen mineralization in eucalypt forests of south-eastern Australia. I. Nutrient cycling and nitrogen turnover. *Plant and Soil* **92**, 319–339.

Adams MA, Ineson P, Binkley D, Cadisch G, Tokuchi N, Scholes M and Hicks K (2004) Excess nitrogen and ecosystem function: toward a global perspective. *Ambio* **33**, 530–536.

Aerts R and Chapin III FS (2000) The mineral nutrition of wild plants revisited: a re-evaluation of processes and patterns. *Advances in Ecological Research* **30**, 1–67.

Ainsworth EA and Long SP (2005) What have we learned from 15 years of free-air CO_2 enrichment (FACE)? A meta-analytic review of the responses of photosynthesis, canopy properties and plant production to rising CO2. *New Phytologist* **165**, 351–372.

Allaby M (1998) *A Dictionary of Ecology.* Oxford University Press, Oxford.

Andersen AN, Cook CD and Williams RJ (2003) *Fire in Tropical Savannas: The Kapalga Experiment.* Springer, New York.

Anderson IC, Bastias BA, Genney DR, Parkin PI and Cairney JWG (2007) Basidiomycete fungal communities in Australian sclerophyll forest soil are altered by repeated prescribed burning. *Mycological Research* **111**, 482–486.

Association for Fire Ecology (2009) The role of fire in managing long-term carbon stores: key challenges. http://fireecology.net.

Attenborough D (1995) *The Private Life of Plants*. BBC Books, London.

Attiwill PM (1979) Nutrient cycling in a *Eucalyptus obliqua* (L' Hérit.) forest. III. Growth, biomass and net primary production. *Australian Journal of Botany* **27**, 439–458.

Attiwill PM (1992) Productivity of *Eucalyptus regnans* forest regenerating after bushfire. *South African Forestry Journal* **160**, 1–6.

Attiwill PM (1994) The disturbance of forest ecosystems and the conservative management of eucalypt forests in Australia. *Forest Ecology and Management* **63**, 301–346.

Attiwill PM and Adams MA (1993) Tansley Review No 50: Nutrient cycling in forests. *New Phytologist* **124**, 561–582.

Attiwill PM and Adams MA (2008) Harnessing forest ecological sciences in the service of stewardship and sustainability: a perspective from 'down-under'. *Forest Ecology and Management* **256**, 1636–1645.

Attiwill PM, England J and Whittaker K (2001) *The Environmental Credentials of Production, Manufacture and Re-use of Wood Fibre in Australia*. Department of Agriculture Fisheries Forestry, Canberra.

Attiwill PM and Leeper GW (1987) *Forest Soils and Nutrient Cycles*. Melbourne University Press, Carlton.

Attiwill PM and May BM (2001) Does nitrogen limit the growth of native eucalypt forests: some observations for mountain ash (*Eucalyptus regnans*). *Marine and Freshwater Research* **52**, 111–117.

Attiwill PM and Packham D (2009) Are big fires inevitable? In *The People's Review of Bushfires, 2002–2007, in Victoria*. (Eds PM Attiwill, D Packham, T Barker and I Hamilton I) pp. 44–51. The People's Review: Richmond, Victoria.

Attiwill PM and Wilson BA (2006) Succession, disturbance, and fire. In *Ecology: An Australian Perspective*. 2nd edn. (Eds PM Attiwill and BA Wilson) pp. 361–377. Oxford University Press, Melbourne.

Attiwill PM, Packham D, Barker T and Hamilton I (Eds) (2009) *The People's Review of Bushfires, 2002–2007, in Victoria*. The People's Review: Richmond, Victoria.

Attiwill PM, Polglase PJ, Weston CJ and Adams MA (1996) Nutrient cycling in forests of south-eastern Australia. In *Nutrition of Eucalypts*. (Eds PM Attiwill and MA Adams) pp. 191–227. CSIRO Publishing, Melbourne.

Australian Centre for Biodiversity, Monash University (2009) Submission to 2009 Victorian Bushfires Royal Commission. http://www.royalcommission.vic.gov.au/Submissions/SubmissionDocuments/SUBM-002-030-0292_R.pdf.

Baldock JA and Smernik RJ (2002) Chemical composition and bioavailability of thermally altered *Pinus resinosa* (Red pine) wood. *Organic Geochemistry* **33**, 1093–1109.

Barrett DJ and Gifford RM (1995) Photosynthetic and growth acclimation to elevated CO_2 in cotton: interactions with severe phosphate deficiency and limited rooting volume. *Australian Journal of Plant Physiology* **22**, 955–963.

Barrett DJ, Richardson AE and Gifford RM (1998) Elevated atmospheric CO_2 concentrations increase wheat root phosphatase activity when growth is limited by phosphorus. *Australian Journal of Plant Physiology* **25**, 87–93.

Barton CVM, Ellsworth DS, Medlyn BE, Tissue DT, Duursma R, Adams MA, Eamus D, Conroy JP, McMurtrie RE, Parsby J and Linder S (2011). Whole-tree chambers for elevated atmospheric CO_2 experimentation and tree-scale flux measurements in south-eastern

Australia: the Hawkesbury Forest Experiment. *Agricultural and Forest Meteorology* (in press).

Bastias B, Huang ZQ, Blumfield T, Xu Z and Cairney JWG (2006) Influence of repeated prescribed burning on the soil fungal community in an eastern Australian wet sclerophyll forest. *Soil Biology and Biochemistry* **38**, 3492–3501.

Bastias BA, Xu Z and Cairney JWB (2006) Influence of long-term repeated prescribed burning on mycelial communities of ectomycorrhizal fungi. *New Phytologist* **172**, 149–158.

Benyon R, Culvenor D, Sims N, Opie K, Siggins A and Doody T (2007) 'Evaluation of remote sensing for predicting long term hydrological impacts of forest regeneration as a result of bushfire'. Technical Report No. 163. Ensis, Canberra.

Binkley D (2005) How nitrogen-fixing trees change soil carbon. In *Tree Species Effects on Soils: Implications for Global Change.* (Eds D Binkley and O Menyailo) pp. 155–164. NATO Science Series, Springer, Dordrecht.

Binkley D, Giardina C and Bashkin M (2000) Soil phosphorus pools and supply under the influence of *Eucalyptus saligna* and nitrogen-fixing *Albizia facaltaria. Forest Ecology and Management* **128**, 241–247.

Biswell HH (1989) *Prescribed Burning in California Wildlands and Vegetation Management.* University of California Press, Berkeley.

Blainey GN (1982) *Triumph of the Nomads.* Macmillan, Melbourne.

Blais JR (1983) Trends in the frequency, extent and severity of spruce budworm outbreaks in eastern Canada. *Canadian Journal of Forest Research* **13**, 539–547.

Boer MM, Sadler, RJ, Wittkuhn RS, McCaw L and Grierson PF (2009) Long-term impacts of prescribed burning on regional extent and incidence of wildfires – evidence from 50 years of active fire management in SW Australian forests. *Forest Ecology and Management* **259**, 132–142.

Boerner REJ (1982) Fire and nutrient cycling in temperate ecosystems. *BioScience* **32**, 187–192.

Bond WJ and van Wilgen BW (1996) *Fire and Plants.* Chapman & Hall, London.

Bond WJ, Woodward FI and Midgley GF (2005) The global distribution of ecosystems in a world without fire. *New Phytologist* **165**, 525–538.

Bormann FH and Likens GE (1979) *Pattern and Process in a Forested Ecosystem.* Springer-Verlag, New York.

Bottrill M, Joseph LN, Carwardine J, Bode M, Cook C, Game ET, Grantham HS, Kark S, Linke S, McDonald-Madden E, Pressey RL, Walker S, Wilson KA and Possingham HP (2009) Finite funds means triage is unavoidable. *Trends in Ecology and Evolution* **23**, 649–654.

Bowman DMJS (1998) Tansley Review No. 101. The impact of Aboriginal landscape burning on the Australian biota. *New Phytologist* **140**, 385–410.

Bowman DMJS (2000) *Australian Rainforests. Islands of Green in a Land of Fire.* Cambridge University Press, Cambridge, UK.

Bowman DMJS (2003a) Australian landscape burning: a continental and evolutionary perspective. In *Fire in Ecosystems of South-west Australia: Impacts and Management.* (Eds I Abbott and N Burrows). pp. 107–118. Backhuys Publishers, Leiden, The Netherlands.

Bowman DMJS (2003b) Wild, tame and feral fire: the fundamental linkage between indigenous fire usage and the conservation of Australian biodiversity. Paper presented to The 3rd International Wildland Fire Conference and Exhibition, Melbourne, Australia.

Bowman DMJS (2005) Understanding a flammable planet – climate, fire and global vegetation patterns. *New Phytologist* **165**, 314–145.

Bowman DMJS, Balch JK, Artaxo P, Bond WJ, Carlson JM, Cochrane MA *et al.* (2009) Fire in the earth system. *Science* **324**, 481–484.

Bowman DMJS, McLean AR and Crowden RK (1986) Vegetation-soil relations in the lowlands of south-west Tasmania. *Australian Journal of Ecology* **11**, 141–153.

Bradstock RA (2008) Effects of large fires on biodiversity in south-eastern Australia: disaster or template for diversity? *International Journal of Wildland Fire* **17**, 809–822.

Bradstock RA, Williams JE and Gill AM (Eds) (2002) *Flammable Australia. The Fire Regimes of a Continent.* Cambridge University Press, Cambridge, UK.

Buckley TN and Roberts DW (2006) DESPOT: a tree growth model that allocates carbon to maximize carbon gain. *Tree Physiology* **26**, 129–144.

Buckley TN (2008) The role of stomatal acclimation in modelling tree adaptation to high CO_2. *Journal of Experimental Botany* **59**, 1951–1961.

Bureau of Meteorology (2009) Changes to the Bureau's fire weather warnings and forecasts. http://www.bom.gov.au/weather-services/bushfire/fire-warnings-oct-2009.shtml.

Burrell JP (1981) The invasion of coastal heaths of Victoria by *Leptospermum laevigatum* (J. Gaerth) F. Muell. *Australian Journal of Botany* **29**, 747–764.

Burrows N (2008) Linking fire ecology and fire management in south-west Australian forest landscapes. *Forest Ecology and Management* **255**, 2394–2406.

Burrows N and Abbott I (2003) Fire in south-west Western Australia: synthesis of current knowledge, management implications and new research directions. In *Fire in Ecosystems of South-Western Australia: Impacts and Management.* (Eds I Abbott and N Burrows) pp. 437–452. Backhuys Publishers, Leiden, The Netherlands.

Campbell A (2008) *Managing Australian Landscapes in a Changing Climate: A Climate Change Primer for Regional Natural Resource Management Bodies.* Department of Climate Change, Canberra. http://www.climatechange.gov.au/~/media/publications/adaptation/managing-australian-landscapes.ashx.

Cary GJ and Bradstock RA (2003) Sensitivity of fire regimes to management. In *Australia Burning: Fire Ecology, Policy and Management Issues.* (Eds G Cary, D Lindenmayer and S Dovers) pp. 65–81. CSIRO Publishing, Melbourne.

Cary GJ and Morrison DA (1995) Effects of fire frequency on plant species composition of sandstone communities in the Sydney region: combinations of inter-fire intervals. *Austral Ecology* **20**, 418–426.

Catling PC and Newsome AE (1981) Responses of the Australian vertebrate fauna to fire: an evolutionary approach. In *Fire and the Australian Biota.* (Eds AM Gill, RH Groves and IR Noble) pp. 273–310. Australian Academy of Science, Canberra.

Certini G (2005) Effects of fire on properties of forest soils: a review. *Oecologia* **143**, 1–10.

Chambers DP and Attiwill PM (1994) The ash-bed effect in *Eucalyptus regnans* forest: chemical, physical and microbiological changes in soil after heating and partial sterilization. *Australian Journal of Botany* **42**, 739–749.

Cheney NP (1981) Fire behaviour. *Fire and the Australian Biota.* (Eds AM Gill, RH Groves and IR Noble) pp. 151–175. Australian Academy of Science, Canberra.

Cheney NP (2003) Effectiveness of prescribed burning on reducing fire behaviour. Paper presented to a Symposium, 'Bush Fire Prevention: Are We Doing Enough?' Institute of Public Affairs, Melbourne.

Cheney NP (2004) 'The bush capital or the bushfire capital?' Canberra Day Oration, 2004.

Christensen NL, Agee JK, Brussard PF, Hughes J, Knight DH, Minshall GW, Peek JM, Pyne SJ, Swanson FJ, Thomas JW, Wells S, Williams SE and Wright HA (1989) Interpreting the Yellowstone fires of 1988. *BioScience* **39**, 678–685.

Centre for International Forest Research (2009) Simply REDD. CIFOR's guide to forests, climate change and REDD. CIFOR, Bogor, Indonesia.

Clack P (2003) *Firestorm. Trial by Fire.* John Wiley & Sons, Australia.

Clements FE (1916) *Plant Succession: An Analysis of the Development of Vegetation.* Carnegie Institute, Washington DC.

Close DC, Davidson NJ, Johnson DW, Abrams MD, Hart SC, Lunt ID, Archibald RD, Horton B and Adams MA (2009) Premature decline of *Eucalyptus* and altered ecosystem processes in the absence of fire in some Australian forests. *The Botanical Review* **75**, 191–202.

Colinvaux P (1993) *Ecology 2.* John Wiley & Sons, New York.

Collins P (2006) *Burn: The Epic Story of Bushfire in Australia.* Allen & Unwin, Sydney.

Commonwealth of Australia (1992) *National Forest Policy Statement.* Australian Government Printing Service, Canberra.

Commonwealth of Australia and Victorian Governments (1998) *Central Highlands Regional Forest Agreement.* The Commonwealth of Australia and The State of Victoria.

Cornish PM and Vertessy RA (2001) Forest age-induced changes in evapotranspiration and water yield from a eucalypt forest. *Journal of Hydrology* **242**, 43–63.

Crawley M J (1997) The structure of plant communities. In *Plant Ecology.* (Ed. MJ Crawley) pp. 475–531. Blackwell Science, Oxford.

Crutzen PJ and Andreae MO (1990) Biomass burning in the tropics: impact on atmospheric chemistry and biogeochemical cycles. *Science* **250**, 1669–1678.

Czimczik CI, Preston CM, Schmidt MWI, Werner RA and Schulze E-D (2002) Effects of charring on mass, organic carbon, and stable carbon isotope composition of wood. *Organic Geochemistry* **33**, 1207–1223.

Czimczik CI, Schmidt MWI and Schulze E-D (2005) Effects of increasing fire frequency on black carbon and organic matter in Podzols of Siberian Scots pine forests. *European Journal of Soil Science* **56**, 417–428.

Dargavel J (1994) Constructing Australia's forests in the image of capital. In *Australian Environmental History. Essays and Cases.* (Ed. S Dovers) pp. 80–98. Oxford University Press, Melbourne.

DeLuca TH and Aplet GH (2008) Charcoal and carbon storage in forest soils of the Rocky Mountain west. *Frontiers in Ecology and the Environment* **6**, 18–24.

Department of Sustainability and Environment (2003) 'Ecological effects of repeated low-intensity fire in mixed eucalypt foothill forest in south-eastern Australia: summary report (1984–1999)'. Fire Research Report No. 57. Department of Sustainability and Environment, Victoria.

Department of Sustainability and Environment, Victoria (2006) Code of Practice for Fire Management on Public Land (Revision 1). Department of Sustainability and Environment, Victoria.

Department of Sustainability and Environment, Victoria (2010). Bushfire statistics – fires on public land in Victoria. http://www.dse.vic.gov.au/DSE/nrenfoe.nsf/fid/50583FE0BE02C9 D0CA2576650081F856.

Department of Sustainability and Environment, Victoria (2010). Major bushfires in Victoria. http://www.dse.vic.gov.au/DSE/nrenfoe.nsf/LinkView/E20ACF3A4A127CB04A2567930 0155B04D79E4FB0C437E1B6CA256DA60008B9EF.

Department of the Environment, Water, Heritage and the Arts (2006) State of the Environment 2006. Indicator: BD-11 Area burnt by frequency, intensity and season of burning. http://www.environment.gov.au/soe/2006/publications/drs/indicator/101/index.html.

Dexter B (2005) *The Facts Behind the Fire: A Scientific and Technical Review of the Circumstances Surrounding the 2003 Victorian Bushfire Crisis.* Forest Fire Vic Inc., Melbourne.

Dixon K and Barrett R (2003) Defining the role of fire in south-west Western Australian plants. In *Fire in Ecosystems of South-west Western Australia*. (Eds I Abbott and N Burrows) pp. 205–223. Backhuys Publishers, Leiden, The Netherlands.

Donato DC, Campbell JL, Fontaine JB and Law BE (2009) Quantifying char in postfire woody detritus inventories. *Fire Ecology* 5, 104–115.

Dowdy AJ, Mills GA, Finkele K and de Groot W (2009) *Australian Weather as Represented by the McArthur Forest Fire Danger Index and the Canadian Forest Fire Weather Index*. Centre for Australian Weather and Climate Research. Bureau of Meteorology and CSIRO, Australia.

Drury WH and ICT Nisbet (1973) Succession. *Journal of the Arnold Arboretum, Harvard University* 54, 331–368.

Enright NJ and Thomas I (2008) Pre-European fire regimes in Australian ecosystems. *Geography-Compass* 2, 979–1011.

Fensham RJ, Fairfax RJ and Archer SR (2005) Rainfall, land use and woody vegetation cover change in semi-arid Australian savanna. *Journal of Ecology* 93, 596–606.

Fernandes PM and Botelho HS (2003) A review of prescribed burning effectiveness in fire hazard reduction. *International Journal of Wildland Fire* 12, 117–128.

Finney MA, McHugh CW and Grenfell IC (2005) Stand- and landscape-level effects of prescribed burning on two Arizona wildfires. *Canadian Journal of Forest Research* 35, 1714–1722.

Flannery TF (1994) *The Future Eaters: An Ecological History of the Australasian Lands and People*. Reed Books, Sydney.

Fontaine S, Barot S, Barre P, Bdioui N, Mary B and Rumpel C (2007) Stability of organic carbon in deep soil layers controlled by fresh carbon supply. *Nature* 450, 277–280.

Fontaine S, Mariotti A and Abbadie L (2003) The priming effect of organic matter: a question of microbial competition? *Soil Biology and Biochemistry* 35, 837–843.

Forbes MS, Raison RJ and Skjemstad JO (2006) Formation, transformation and transport of black carbon (charcoal) in terrestrial and aquatic ecosystems. *Science of the Total Environment* 370, 190–206.

Force DC (1981) Postfire insect succession in southern California chaparral. *The American Naturalist* 117, 575–582.

Fox BJ (1983) Mammal species diversity in Australian heathlands: the importance of pyric succession and habitat diversity. In *Mediterranean-type Ecosystems: The Role of Nutrients*. (Eds FJ Kruger, DT Mitchell and JUM Jarvis) pp. 473–489. Springer-Verlag, Berlin.

Friend GR (1999) Fire and faunal response patterns – a summary of research findings. In *Management of Fire for the Conservation of Biodiversity – Workshop Proceedings*. (Eds G Friend, M Leonard, A MacLean and I Sieler) pp. 39–45. Department of Natural Resources and Environment, Melbourne.

Friend GR and Wayne A (2003) Relationships between mammals and fire in south-west Western Australian ecosystems: what we know and what we need to know. In *Fire in Ecosystems of South-west Australia: Impacts and Management*. (Eds I Abbott and N Burrows) pp. 363–380. Backhuys Publishers, Leiden, The Netherlands.

Gaston KJ (2010) Valuing common species. *Science* 327, 154–155.

Gill AM (1981) Adaptive responses of Australian vascular plant species to fires. In *Fire and the Australian Biota*. (Eds AM Gill, RH Groves and IR Noble) pp. 243–272. Australian Academy of Science, Canberra.

Gill AM (2008) 'Underpinnings of fire management for biodiversity in reserves. Fire and adaptive management'. Report No. 73. Victorian Government Department of Sustainability and Environment, Melbourne.

Gill AM and Allan G (2008) Large fires, fire effects and the fire-regime concept. *International Journal of Wildland Fire* 17, 688–695.

Gill AM and Bradstock RA (1992) A national register for the fire responses of plant species. *Cunninghamia* 2, 653–660.

Gill AM and McCarthy MA (1998) Intervals between prescribed fires in Australia: what intrinsic variation should apply? *Biological Conservation* 85, 161–169.

Gill AM and Moore PHR (1997) 'Contemporary fire regimes in the forests of southwestern Australia'. Contract Report to Environment Australia, CSIRO Plant Industry, Canberra.

Gould JS, McCaw WL, Cheney NP, Ellis PF, Knight IK and Sullivan AL (2007) *Project Vesta – Fire in Dry Eucalypt Forest: Fuel Structure, Fuel Dynamics and Fire Behaviour.* ENSIS-CSIRO, Canberra, and Department of Environment and Conservation, Perth.

Guenther AB, Hewitt CN, Erickson D, Fall R, Geron C, Graedel T *et al.* (1995) A global model of natural volatile organic compound emissions. *Journal of Geophysical Research* 100, 8873–8892.

Hammes K, Schmidt MWI, Smernik RJ, Currie LA, Ball WP, Nguyen TH *et al.* (2007) Comparison of quantification methods to measure fire-derived (black/elemental) carbon in soils and sediments using reference materials from soil, water, sediment and the atmosphere. *Global Biogeochemical Cycles* 21, 1–18.

Hamilton SD, Lawrie AC, Hopmans P and Leonard BV (1991) Effects of fuel reduction burning on a *Eucalyptus obliqua* forest ecosystem in Victoria. *Australian Journal of Botany* 39, 203–217.

Han W, Fang J, Guo D and Zhang Y (2005) Leaf nitrogen and phosphorus stoichiometry across 753 terrestrial plant species in China. *New Phytologist* 168, 377–385.

He C, Murray F and Lyons T (2000) Monoterpene and isoprene emissions from 15 *Eucalyptus* species in Australia. *Atmospheric Environment* 34, 645–655.

Hedin LO (2004) Global organization of terrestrial plant-nutrient interactions. *Proceedings of the National Academy of Sciences* 101, 10849–10850.

Hill RS (2004) Origins of southeastern Australian vegetation. *Philosophical Transactions of the Royal Society of London, B* 359, 1537–1549.

Hodgson A (1968) Control burning in eucalypt forests in Victoria, Australia. *Journal of Forestry* 66, 601–605.

Hodgson A (2004) The feral fire. Paper presented to *Bushfire, the bush's salvation*, a Public Meeting at Bairnsdale, Victoria, 23 May 2004.

Hoggett J and Hoggett A (2004) When will we ever learn? IPA Backgrounder 16/2. Institute of Public Affairs Ltd, Melbourne.

Hopmans P, Bauhus J, Khanna P and Weston C (2005) Carbon and nitrogen in forest soils: Potential indicators for sustainable management of eucalypt forests in south-eastern Australia. *Forest Ecology and Management* 220, 75–87.

Hughes L (2003) Climate change and Australia: trends, projections and impacts. *Austral Ecology* 28, 423–443.

Humphreys FR and Craig FG (1981) Effects of fire on soil chemical, structural and hydrological properties. In *Fire and the Australian Biota*. (Eds AM Gill, RH Groves and IR Noble) pp. 177–200. Australian Academy of Science, Canberra.

Humphries RK (1994) The effects of single autumn and spring prescribed fires on small mammal and reptile ecology in Wombat State Forest. Masters Thesis in Applied Science. University of Ballarat, Australia.

Hungate BA, Dukes JS, Shaw MR, Luo Y and Field CB (2003a) Nitrogen and climate change. *Science* **302**, 1512–1513.

Hungate BA, Naiman RJ, Apps M, Cole JJ, Moldan B, Satake K, Stewart JWB, Victoria R and Vitousek PM (2003b) Disturbance and element interactions. In *Interactions of the Major Biogeochemical Cycles, Global Change and Human Impacts*. (Eds JM Melillo, CB Field and B Moldan) pp. 47–62. Island Press, Washington DC.

IPCC (2001) *Climate Change 2001: The Scientific Basis. Contribution of Working Group 1 to the Third Assessment Report of the Intergovernmental Panel on Climate Change*. Cambridge University Press, Cambridge, UK.

Jackson WD (2000) Nutrient stocks in Tasmanian vegetation and approximate losses due to fire. *Papers and Proceedings of the Royal Society of Tasmania* **134**, 1–18.

Jenkins ME and Adams MA (2010) Vegetation type determines heterotrophic respiration in sub-alpine Australian ecosystems. *Global Change Biology* **16**, 209–219.

Jurskis V (2000) Vegetation changes since European settlement of Australia: an attempt to clear up some burning issues. *Australian Forestry* **63**, 166–173.

Jurskis V (2003) Assessing the ecological implications of prescribed burning: where do you start? *Third International Wildland Fire Conference and Exhibition, Sydney*. Australasian Fire Authorities Council, Melbourne.

Jurskis V (2005a) Decline of eucalypt forests as a consequence of unnatural fire regimes. *Australian Forestry* **68**, 257–262.

Jurskis V (2005b) Eucalypt decline in Australia, and a general concept of tree decline and dieback. *Forest Ecology and Management* **215**, 1–20.

Jurskis V, Bridges B and de Mar P (2003) Fire management in Australia: the lessons of 200 years. In *Joint Australia and New Zealand Institute of Forestry Conference Proceedings 27 April–1 May 2003, Queenstown, New Zealand*. pp. 353–368. Ministry of Agriculture and Forestry, Wellington.

Jurskis V and Turner J (2002) Eucalypt decline in eastern Australia: a simple model. *Australian Forestry* **65**, 81–92.

Kasel S (1999) The decline of *Eucalyptus camphora* and *E. ovata* within the Yellingbo State Nature Reserve, Victoria. PhD Thesis, University of Melbourne.

Keay J and Bettenay E (1969) Concentrations of major nutrient elements in vegetation from a portion of the Western Australian arid zone. *Journal of the Royal Society of Western Australia* **52**, 109–118.

Kenny B, Sutherland E, Tasker E and Bradstock B (2004) *Guidelines for Ecologically Sustainable Fire Management*. NSW National Parks and Wildlife Service, Hurstville, NSW.

Kerkhoff AJ and Enquist BJ (2006) Ecosystem allometry: the scaling of nutrient stocks and primary productivity across plant communities. *Ecology Letters* **9**, 419–427.

Keith H, Mackey BG and Lindenmayer DB (2009) Re-evaluation of forest biomass carbon stocks and lessons from the world's most carbon-dense forests. *Proceedings of the National Academy of Science, USA* **106**, 11635–11640.

King DJ, Gleadow RM and Woodrow IE (2004) Terpene deployment in *Eucalyptus polybractea*; relationships with leaf structure, environmental stresses and growth. *Functional Plant Biology* **31**, 451–460.

King KJ, Cary GJ, Bradstock RA, Chapman, J, Pyrke D and Marsden-Smedley JB (2006) Simulation of prescribed burning strategies in south-west Tasmania, Australia: effects on unplanned fires, fire regimes, and ecological management values. *International Journal of Wildland Fire* **15**, 527–540.

Kirschbaum MUF (2000) Will changes in soil organic matter act as a positive or negative feedback on global warming? *Biogeochemistry* **48**, 21–51.

Koerselman W and Meuleman AFM (1996) The vegetation N:P ratio: a new tool to detect the nature of nutrient limitation. *Journal of Applied Ecology* **33**, 1441–1450.

Kruger FJ (1983) Plant community diversity and dynamics in relation to fire. In *Mediterranean-type Ecosystems: The Role of Nutrients*. (Eds FJ Kruger, DT Mitchell and JUM Jarvis) pp. 446–472. Springer-Verlag, Berlin.

Krull E, Baldock J, Smernik R and Skjemstad J (2009) Characterization of biochar. In *Biochar for Environmental Management: Science and Technology*. (Eds J Lehmann and S Joseph) pp. 53–66. Earthscan, UK.

Krull E, Lehmann J, Skjemstad J and Baldock J (2008) The global extent of Black C in soils: is it everywhere? In *Grasslands: Ecology, Management and Restoration*. (Ed. H Schröder) pp. 13–17. Nova Publishers, Hauppauge NY.

Kuczera G (1987) Prediction of water yield reductions following a bushfire in ash-mixed species eucalypt forest. *Journal of Hydrology* **94**, 215–236.

Lane PNJ and MacKay SM (2001) Streamflow response of mixed-species eucalypt forests to patch cutting and thinning treatments. *Forest Ecology and Management* **143**, 131–142.

Lehmann J and Sohi S (2008) Comment on 'Fire-derived charcoal causes loss of forest humus'. *Science* **321**, 1295.

Lehmann J, Skjemstad J, Sohi S, Carter J, Barson M, Falloon P, *Coleman K, Woodbury P and Krull E* (2008) Australian climate-carbon cycle feedback reduced by soil black carbon. *Nature Geoscience* **1**, 832–835.

Liang B, Lehmann J, Solomon D, Kinyangi J, Grossman J, O'Neill B, Skjemstad JO, Thies J, Luizão FJ, Petersen J and Neves EG (2006) Black Carbon increases cation exchange capacity in soils. *Soil Science Society of America Journal* **70**, 1719–1730.

Liang B, Lehmann J, Solomon D, Sohi S, Thies JE, Skjemstad JO, Luizão FJ, Engelhard MH, Neves EG and Wirick S (2008) Stability of biomass-derived black carbon in soils. *Geochimica et Cosmochimica Acta* **72**, 6069–6078.

Lloyd J, Bird MI, Veenendaal EM and Kruijt B (2001) Should phosphorus availability be constraining moist tropical forest response to increasing CO_2 concentrations? In *Global Biogeochemical Cycles in the Climate System*. (Eds E-D Schulze, M Heinman, S Harrison, E Holland, J Lloyd, IC Prentice and D Schimel) pp. 95–114. Academic, San Diego.

Loucks OL (1970) Evolution of diversity, efficiency, and community stability. *American Zoologist* **10**, 17–25.

Loreto F and Delfine S (2000) Emission of isoprene from salt-stressed *Eucalyptus globulus* leaves. *Plant Physiology* **123**, 1605–1610.

Lucas C, Hennessy K, Mills G and Bathols J (2007) 'Bushfire weather in southeast Australia: recent trends and projected climate change impacts'. Consultancy report prepared for The Climate Institute of Australia. Bushfire Cooperative Research Centre, Melbourne.

Luke RH and McArthur AG (1978) *Bushfires in Australia*. Australian Government Publishing Office, Canberra.

Macfarlane C, Bond C, White DA, Grigg A, Ogden GN and Siberstein R (2010) Transpiration and hydraulic traits of old and regrowth eucalypt forest in southwestern Australia. *Forest Ecology and Management* **260**, 96–105.

Mackey B, Lindenmeyer D, Gill M, McCarthy M and Lindsay J (2002) *Wildlife, Fire and Future Climate*. CSIRO Publishing, Melbourne.

Maleknia S, Vail TM, Cody RB, Sparkman DO, Bell TL and Adams MA (2008) Temperature-dependent analysis of volatile organic compounds of eucalypts by Direct Analysis in Real Time (DART) mass spectrometry. *Rapid Communications in Mass Spectroscopy* **23**, 2241–2246.

Maleknia SD, Bell TL and Adams MA (2009) Eucalypt smoke and wildfires: temperature dependent emissions of biogenic volatile organic compounds. *International Journal of Mass Spectrometry* **279**, 126–133.

Marris E (2006) Black is the new green. *Nature* **442**, 624–626.

Matthews S (2006) A process-based model of fine fuel moisture. *International Journal of Wildland Fire* **15**, 155–168.

May B and PM Attiwill (2003) Nitrogen-fixation by *Acacia dealbata* and changes in soil properties 5 years after mechanical disturbance or slash-burning following timber harvest. *Forest Ecology and Management* **217**, 1–17.

McArthur AG (1962) 'Control burning in eucalypt forest'. Leaflet No. 80. Australian Forestry and Timber Bureau. Canberra.

McArthur AG (1967) 'Fire behaviour in eucalypt forests'. Leaflet 107. Forestry and Timber Bureau, Canberra.

McCarthy GJ and Tolhurst KG (1998) 'Effectiveness of fire-fighting first attack operations, NRE Victoria 1991/92–1994/95'. Research Report No. 45. Fire Management. Department of Natural Resources and Environment, Victoria.

McCarthy GJ and Tolhurst KG (2001) 'Effectiveness of broadscale fuel reduction burning in assisting with wildfire control in parks and forests in Victoria'. Research Report 51. Fire Management, Department of Natural Resources and Environment, Victoria.

McCarthy GJ, Tolhurst KG and Chatto K (2003) 'Determination of sustainable fire regimes in the Victorian Alps using plant vital attributes'. Research Report No. 54. Department of Sustainability and Environment, Victoria.

McCarthy GJ, Tolhurst KG and Chatto K (2009) 'Overall fuel hazard'. Research Report No. 47, 3rd edn. Fire Management, Department of Sustainability and Environment, Victoria.

McCaw WL, Gould JS and Cheney NP (2008) Quantifying the effectiveness of fuel management in modifying wildfire behaviour. *International Bushfire Research Conference 2008 incorporating the 15th Annual AFAC Conference, Adelaide.*

McDonald-Madden E, Baxter PWJ and Possingham HP (2008a) Making robust decisions for conservation with restricted money and knowledge. *Journal of Applied Ecology* **45**, 1630–1638.

McDonald-Madden E, Baxter PWJ and Possingham HP (2008b) Subpopulation triage: how to allocate conservation effort among populations. *Conservation Biology* **22**, 656–665.

McDougall KL (2003) Aerial photographic interpretation of vegetation changes on the Bogong High Plains, Victoria, between 1936 and 1980. *Australian Journal of Botany* **51**, 251–256.

McGroddy ME, Daufresne T and Hedin LO (2004) Scaling of C:N:P stoichiometry in forests worldwide: implications of terrestrial Redfield-type ratios. *Ecology* **85**, 2390–2401.

McIntosh PD, Laffan MD and Hewitt AE (2005) The role of fire and nutrient loss in the genesis of the forest soils of Tasmania and southern New Zealand. *Forest Ecology and Management* **220**, 185–215.

Melillo JM, Field CB and Molden B (Eds) (2003) SCOPE 61. *Interaction of the Major Biogeochemical Cycles: Global Change and Human Impacts.* Island Press, Washington DC.

Mooney SD, Radford KL and Hancock G (2001) Clues to the 'burning question': pre-European fire in the Sydney coastal region from sedimentary charcoal and palynology. *Ecological Management and Restoration* **2**, 203–212.

Morrison DA, Cary GJ, Pengelly SM, Ross DG, Mullins BJ, Thomas CR and Anderson TS (1995) Effects of fire frequency on plant species composition of sandstone communities in the Sydney region: Inter-fire interval and time-since-fire. *Austral Ecology* **20**, 239–247.

Mutch RW (1970) Wildland fires and ecosystems – a hypothesis. *Ecology* **51**, 1046–1051.

Nature Conservation Council (2010) http://nccnsw.org.au/index.php?option=com_content&task=blogcategory&id=349&Itemid=691.

Neumann F and Tolhurst K (1991) Effects of fuel reduction burning on epigeal arthropods and earthworms in dry sclerophyll forest of west-central Victoria. *Australian Journal of Ecology* **16**, 315–330.

Newsome AE, McIlroy J and Catling P (1975) The effects of an extensive wildfire on populations of 20 ground vertebrates in south-east Australia. *Proceedings of the Ecological Society of Australia* **9**, 107–128.

New South Wales Department of Environment, Climate Change and Water (2010) http://www.environment.nsw.gov.au/firemanagement/BlueMountainsNPfms.htm.

Nicholson PH (1981) Fire and the Australian Aborigine – an enigma. In *Fire and the Australian Biota*. (Eds AM Gill, RH Groves and IR Noble) pp. 55–76. Australian Academy of Science, Canberra.

Noble WS (1977) *Ordeal by Fire. The Week a State Burned Up.* Jenkin Buxton, Melbourne.

NSW National Parks and Wildlife Service (2002) *New South Wales Flora Fire Response Database Version 1.3.* (Ed. J Cohn). Biodiversity Research & Management Division, NSW National Parks and Wildlife Service, Hurstville.

O'Connell AM and Mendham DS (2004) Impact of N and P fertilizer application on nutrient cycling in jarrah (*Eucalyptus marginata*) forests of south western Australia. *Biology and Fertility of Soils* **40**, 136–143.

Osborn M, McLean CB, Burns SC and Bell TL (2003) Effects of fire on the mycorrhizal diversity of *Caladenia formosa* (Orchidaceae). International Conference on Mycorrhizas 4, Montreal, Canada.

Ostaff DP and MacLean DA (1989) Spruce budworm populations, defoliation, and changes in stand condition during an uncontrolled spruce budworm outbreak on Cape Breton Island, Nova Scotia. *Canadian Journal of Forest Research* **19**, 1077–1086.

Packham D and Attiwill PM (2009) The fire behaviour case for prescribed burning. In *The People's Review of Bushfires, 2002–2007, in Victoria, Final Report.* (Eds P Attiwill, D Packham, T Barker and I Hamilton) pp. 52–58. The People's Review, Richmond.

Page SE, Siegert F, Rieley JO, Boehm H-DV, Jaya A and Limin S (2002) The amount of carbon released from peat and forest fires in Indonesia during 1997. *Nature* **420**, 61–65.

Parsons M, Gavran M and Davidson J (2006) Australia's Plantations 2006. Department of Agriculture, Fisheries and Forestry, Canberra.

Pekin BK, Boer MM, Macfarlane C and Grierson PF (2009) Structural and physiological responses of eucalypt forest to variation in fire frequency and aridity in southwest Australia. *Forest Ecology and Management* **258**, 2136 – 2142.

Penman TD, Kavanagh RP, Binns DL and Melick DR (2007) Patchiness of prescribed burns in dry sclerophyll eucalypt forests in South-eastern Australia. *Forest Ecology and Management* **252**, 24–32.

Penman TD, Binns DL and Kavanagh RP (2008) Quantifying successional changes in response to forest disturbances. *Applied Vegetation Science* **11**, 261–268.

Penman TD, Binns DL, Brasil TE, Shiels RJ and Allen RM (2009) Long-term changes in understorey vegetation in the absence of wildfire in south-east dry sclerophyll forests. *Australian Journal of Botany* **57**, 533–540.

Peters RH (1991) *A Critique for Ecology.* Cambridge University Press, Cambridge, UK.

Pfautsch S, Rennenberg H, Bell TL and Adams MA (2009) Nitrogen uptake by *Eucalyptus regnans* and *Acacia* spp. – preferences, resource overlap and energetic costs. *Tree Physiology* **29**, 389–399.

Pfautsch S, Rennenberg H, Gessler A and Adams MA (2010a) ∂^2H, $\partial^{13}C$ and $\partial^{18}O$ of ecosystem samples identify continental and local climatic influences on hydrology of eucalypt-*Nothofagus* ecosystems. *Water Resources Research* **46**, W03510, doi:10.1029/2009WR007807.

Pfautsch S, Bleby TM, Rennenberg H and Adams MA (2010b) Long-term sap flow measurements reveal influence of temperature and stand structure on water use of *Eucalyptus regnans* forests. *Forest Ecology and Management* **259**, 1190–1199.

Pitman AJ, Narisma GT and McAneney (2007) The impact of climate change on the risk of forest and grassland fires in Australia. *Climate Change* **84**, 383–401.

Prentice IC, Farquhar GD, Fasham MJR, Goulden ML, Heimann M, Jaramillo VJ, Kheshgi HS, Le Quéré C, Scholes RJ and Wallace DWR (2001) The carbon cycle and atmospheric carbon dioxide. In *Climate Change 2001: The Scientific Basis. Contribution of Working Group I to the Third Assessment Report of the Intergovernmental Panel on Climate Change (IPCC).* (Eds JT Houghton, Y Ding, DJ Griggs, M Noguer, PJ van der Linden, X Dai, K Maskell and CA Johnson). pp. 183–237. Cambridge University Press, Cambridge, UK.

Preston CM and Schmidt MWI (2006) Black (pyrogenic) carbon: a synthesis of current knowledge and uncertainties with special consideration of boreal regions. *Biogeosciences Discussions* **3**, 397–420.

Prober SM, Thiele KR and Lunt ID (2007) Fire frequency regulates tussock grass composition, structure and resilience in endangered temperate woodlands. *Austral Ecology* **32**, 808–824.

Pyne SJ (1991) *Burning Bush. A Fire History of Australia.* Henry Holt and Company Inc., New York.

Pyne SJ (2003) Introduction – Fire's Lucky Country. In *Fire in Ecosystems of South-Western Australia: Impacts and Management.* (Eds I Abbott and N Burrows) pp. 1–8. Backhuys Publishers, Leiden, The Netherlands.

Pyne SJ (2007) Problems, paradoxes, paradigms: triangulating fire research. *International Journal of Wildland Fire* **16**, 271–276.

Raison RJ (1980) A review of the role of fire in nutrient cycling in Australian native forests and of methodology for studying the fire-nutrient interaction. *Australian Journal of Ecology* **5**, 15–21.

Raison RJ, Khanna PK and Woods PV (1985a) Mechanisms of element transfer to the atmosphere during vegetation fires. *Canadian Journal of Forest Research* **15**, 132–140.

Raison RJ, Khanna PK and Woods PV (1985b) Transfer of elements to the atmosphere during low-intensity prescribed fires in three Australian sub-alpine eucalypt forests. *Canadian Journal of Forest Research* **15**, 657–664.

Rawson RP, Billing PR and Duncan SF (1983) The 1982–83 forest fires in Victoria. *Australian Forestry* **46**, 163–172.

Rawson R, Billing P and Rees B (1985) 'Effectiveness of fuel reduction burning (10 case studies)'. Research Report No. 25. Department of Sustainability and Environment Victoria.

Redfield AC (1958) The biological control of chemical factors in the environment. *American Scientist* **46**, 205–221.

Reich PB and Oleksyn J (2004) Global patterns of plant leaf N and P in relation to temperature and latitude. *Proceedings of the National Academy of Science, USA* **101**, 11001–11006.

Resh S, Binkley D and Parrotta J (2002) Greater soil carbon sequestration under nitrogen fixing trees when compared with *Eucalyptus* species. *Ecosystems* **5**, 217–231.

Richardson SJ, Peltzer DA, Allen RB, McGlone MS and Parfitt RL (2004) Rapid development of phosphorus limitation in temperate rainforest along the Franz Josef soil chronosequence. *Oecologia* **139**, 267–276.

Richardson SJ, Peltzer DA, Allen RB and McGlone MS (2005) Resorption proficiency along a chronosequence: responses among communities and within species. *Ecology* **86**, 20–25.

Roberts S, Vertessy R and Grayson R (2001) Transpiration from *Eucalyptus sieberi* (L. Johnson) forests of different age. *Forest Ecology and Management* **143**, 153–161.

Romsey Australia. Summary of major bushfires since 1851. http://home.iprimus.com.au/foo7/firesum.html.

Rowe JS and Scotter GW (1973) Fire in the boreal forest. *Quaternary Research (NY)* **3**, 444–464.

Ryan DG, Ryan JE and Starr BJ (undated) *The Australian Landscape – Observations of Explorers and Early Settlers*. Murrumbidgee Catchment Management Committee, Wagga Wagga, New South Wales.

Sardans J and Peñuelas J (2004) Increasing drought decreases phosphorus availability in an evergreen Mediterranean forest. *Plant and Soil* **267**, 367–377.

Schiermeier Q (2005) That sinking feeling. *Nature* **435**, 732–733.

Schimel D and Baker D (2002) The wildfire factor. *Nature* **420**, 29–30.

Schlesinger WH (1997) *Biogeochemistry: An Analysis of Global Change*. 2nd edn. Academic Press, San Diego.

Schmidt MWI, Skjemstad JO, Czimczik CI, Glaser B, Prentice KM, Gelinas Y and Kuhlbusch TAJ (2001) Comparative analysis of black carbon in soils. *Global Biogeochemical Cycles* **15**, 163–167.

Schule W (1990) Landscapes and climate in prehistory: interactions of wildlife, man and fire. In *Fire in the Tropical Biota. Ecological Studies 84*. (Ed. JG Goldammer) pp. 273–318. Springer-Verlag, Berlin.

Schulze E-D, Heinman M, Harrison S, Holland E, Lloyd J, Prentice IC and Schimel D (Eds) (2001) *Global Biogeochemical Cycles in the Climate System*. Academic Press, San Diego.

Shugart HH and West DC (1981) Long term dynamics of forest ecosystems. *American Scientist* **69**, 647–652.

Singh G and Geissler EA (1985) Late Cenozoic history of vegetation, fire, lake levels and climate at Lake George, New South Wales, Australia. *Philosophical Transactions of the Royal Society, London, Series B* **311**, 379–447.

Singh G, Kershaw AP and Clark R (1981) Quaternary vegetation and fire history in Australia. In *Fire and the Australian Biota*. (Eds AM Gill, RH Groves and IR Noble) pp. 23–54. Academy of Science, Canberra.

Skjemstad J and Baldock JA (2007) Total and organic carbon. In *Soil Sampling and Methods of Analysis*. (Ed. MR Carter and EG Gregorich) pp. 225–238. Soil Science Society of Canada, CRC Press, Taylor & Francis Group, Boca Raton, Florida.

Smith NJH (1999) *The Amazon River Forest: A Natural History of Plants, Animals and People*. Oxford University Press, New York.

Sneeuwjagt RJ (2008) Prescribed burning: how effective is it in the control of large bushfires? In *Fire, Environment and Society: From Research to Practice*. pp. 419–435. Bushfire Cooperative Research Centre and the Australasian Fire and Emergency Service Authorities Council, Adelaide.

Sohi S, Lopez-Capel E, Krull E and Bol R (2009) 'Biochar, climate change and soil: a review to guide future research'. CSIRO Land and Water Science Report 05/09. February 2009.

Solomon D, Lehmann J, Thies J, Schäfer T, Liang B, Kinyangi J, Neves E, Petersen J, Luizão F and Skjemstad J (2007) Molecular signature and sources of biochemical recalcitrance of organic C in Amazon Dark Earths. *Geochimica et Cosmochimica Acta* **71**, 2285–2298.

Specht RL and Moll EJ (1983) Mediterranean-type heathlands and sclerophyllous shrublands of the world: an overview. In *Mediterranean-type Ecosystems: The Role of Nutrients*. (Eds FJ Kruger, DT Mitchell and JUM Jarvis) pp. 41–65. Springer-Verlag, Berlin.

Spurr SH and Barnes BV (1980) *Forest Ecology*. 3rd edn. John Wiley & Sons, New York.

State Government of Victoria (2003) 'Report of the Inquiry into the 2002–2003 Victorian Bushfires'. McLaren Press, Abbotsford, Victoria.

Stephens SL Moghaddas JJ, Hartsough BR, Moghaddas EEY and Clinton NE (2009) Fuel treatment effects on stand-level carbon pools, treatment-related emissions, and fire risk in Sierra Nevada mixed-conifer forest. *Canadian Journal of Forest Research* **39**, 1538–1547.

Stewart HTL, Flinn DW and Hopmans P (1985) On harvesting and site productivity in eucalypt forests. *Search* **16**, 206–208.

Stone EL (1979) Nutrient removals by intensive harvest – some research gaps and opportunities. In *Proceedings, Impact of Intensive Harvesting on Forest Nutrient Cycling*. pp. 366–386. College of Environmental Science and Forestry, State University of New York, Syracuse, New York.

Street RA, Hewitt CN and Mennicken S (1997) Isoprene and monoterpene emissions from *Eucalyptus* plantation in Portugal. *Journal of Geophysical Research* **102**, 15875–15888.

Stretton LEB (1939) 'Report of the Royal Commission to inquire into the causes and measures taken to prevent the bush fires of January, 1939, and to protect life and property, and the measures to be taken to prevent bush fires in Victoria and to protect life and property in the event of future bush fires'. Government Printer, Melbourne.

Taranto MT (2003) Relationships among plant communities and underlying soil and water conditions in the Alcoa Lease Area, Anglesea, Victoria. PhD Thesis, University of Melbourne.

Taylor SG (1990) Naturalness: the concept and its application to Australian ecosystems. In *Australian Ecosystems: 200 Years of Utilization, Degradation and Reconstruction*. (Eds D Saunders, AJM Hopkins and RA How). *Proceedings of the Ecological Society of Australia* **16**, 411–418. Surrey Beatty & Sons, Chipping Norton, New South Wales.

Tolhurst KG (1996) Effects of fuel reduction burning on fauna in a dry sclerophyll forest. In *Fire and Biodiversity: The Effects and Effectiveness of Fire Management*. Biodiversity Series, Paper No. 8, pp. 113–121. Biodiversity Unit, Department Environment Sport and Territories, Canberra.

Tolhurst KG (2003) 'Prescribed burning in Victoria: policy and practice.' Paper presented to a symposium, 'Bush Fire Prevention: Are We Doing Enough?' Institute of Public Affairs, Melbourne, 11 March 2003.

Tolhurst KG and Cheney NP (1999) *Synopsis of the Knowledge Used in Prescribed Burning in Victoria*. Department of Natural Resources and Environment, Fire Management, Melbourne.

Underwood RJ, Sneeuwjagt RJ and Styles HG (1985) The contribution of prescribed burning to forest fire control in Western Australia: case studies. In *Fire Ecology and Management of Western Australian Ecosystems*. (Ed. JR Ford) pp. 153–170. WAIT Environmental Studies Group Report No.14. Western Australian Institute of Technology, Perth.

United States Department of Agriculture, Forest Service Southern Region (1989) Technical Publication R8-TP 11. http://www.bugwood.org/pfire/.

Vertessy RA, Watson FR and O'Sullivan SK (2001) Factors determining relations between stand age and catchment water balance in mountain ash forests. *Forest Ecology and Management* **143**, 13–26.

Victorian National Parks Association (2009) Preliminary Submission to the 2009 Victorian Bushfires Royal Commission. http://www.royalcommission.vic.gov.au/Submissions/SubmissionDocuments/SUBM-002-031-0037_R.pdf.

Vitousek PM (2004) *Nutrient Cycling and Limitation: Hawaii as a Model System.* Princeton University Press, Princeton, New Jersey.

Vitousek PM and Howarth RW (1991) Nitrogen limitation on land and in the sea: how can it occur? *Biogeochemistry* **13**, 87–115.

Vivian LM, Doherty MD and Cary GJ (2010) Classifying the fire-response traits of plants: how reliable are species-level classifications? *Austral Ecology* **35**, 264–273.

Walker D (1982) The development of resilience in burned vegetation. In *The Plant Community as a Working Mechanism.* (Ed. EI Newman) pp. 27–43. Blackwell Scientific Publications, Oxford.

Walker J (1981) Fuel dynamics in Australian vegetation. In *Fire and the Australian Biota.* (Eds AM Gill, RH Groves and IR Noble) pp. 101–127. Australian Academy of Science, Canberra.

Walker TW and Syers JK (1976) The fate of phosphorus during pedogenesis. *Geoderma* **15**, 1–19.

Wan S, Hui D and Luo Y (2001) Fire effects on nitrogen pools and dynamics in terrestrial ecosystems: a meta-analysis. *Ecological Applications* **11**, 1349–1365.

Wardell-Johnson G and Burrows N (2004) Towards fire management for a conservation culture: innovations for protecting biodiversity assets in the tingle mosaic, south western-Australia. In *Bushfire in a Changing Environment: New Directions in Management: Proceedings of the 2004 NCC Conference on Ecologically Sustainable Bushfire Management.* pp. 162–167. Nature Conservation Council of New South Wales, Sydney.

Wardle DA, Nilsson M-C and Zackrisson O (2008) Fire-derived charcoal causes loss of forest humus. *Science* **320**, 629.

Wardle DA, Waker LR and Bardgett RD (2004) Ecosystem properties and forest decline in contrasting long-term chronosequences. *Science* **305**, 509–513.

Wassen MJ, Venterink HO, Lapshina ED and Tanneberger F (2005) Endangered plants persist under phosphorus limitation. *Nature* **437**, 547–550.

Watson P (2006) Fire frequency guidelines and the vegetation of the Northern Rivers region. Nature Conservation Council of NSW, Newtown.

Watson PJ, Bradstock RA and Morris EC (2009) Fire frequency influences composition and structure of the shrub layer in an Australian sub-coastal temperate grassy woodland. *Austral Ecology* **34**, 218–232.

Winters AJ, Adams MA, Bleby TM, Rennenberg H, Steigner D, Steinbrecher R and Kreuzwieser J (2009) Emissions of isoprene, monoterpene and short-chained carbonyl compounds from *Eucalyptus* spp. in southern Australia. *Atmospheric Environment* **43**, 3035–3043.

Wood SA, Beringer J, Hutley LB, McGuire AD, Van Dijk A and Kilinc M (2008) Impacts of fire on forest age and runoff in mountain ash forests. *Functional Plant Biology* **35**, 483–492.

Woodward FI (2002) Potential impacts of global elevated CO_2 concentrations on plants. *Current Opinion in Plant Biology* **5**, 207–211.

Woodward FI and Kelly CK (1995) The influence of CO_2 concentration on stomatal density. *New Phytologist* **131**, 311–327.

Whitehead D and Beadle CL (2004) Physiological regulation of productivity and water use in *Eucalyptus*: a review. *Forest Ecology and Management* **193**, 113–140.

York A and Tarnawski J (2004) Impacts of grazing and burning on terrestrial invertebrate assemblages in dry eucalypt forests of north-eastern New South Wales: implications for biodiversity conservation. In *Conservation of Australia's Forest Fauna*. 2nd edn. (Ed. D Lunney) pp. 845–859. Royal Zoological Society of NSW, Mosman.

Zhang L, Hickel K, Dawes WR, Chiew FHS, Western AW and Briggs PR (2004) A rational function approach for estimating mean annual evapotranspiration. *Water Resources Research* **40**, W02502, doi: 10.1029/2003WR002710.

Index

www.ingramcontent.com/pod-product-compliance
Lightning Source LLC
Chambersburg PA
CBHW041130280526
45792CB00013B/2368

* 9 7 8 0 6 4 3 0 9 4 4 3 7 *